# BUILDING MAINTENANCE

**by Jules Oravetz, Sr.**
*Registered Professional Engineer*

THEODORE AUDEL & CO.
*a division of*

HOWARD W. SAMS & CO., INC.
4300 West 62nd Street
Indianapolis, Indiana 46206

SECOND EDITION

SECOND PRINTING—1979

International Standard Book Number: 0-672-23278-2
Library of Congress Catalog Card Number: 76-45885

# Foreword

The maintenance of modern office, industrial, educational, and public buildings is a complex operation calling for many different skills and methods if the job is to be accomplished efficiently. The scope of the subject is so broad that a complete discussion of every phase of building maintenance is impossible. However, in this book the guiding principles are presented in a practical manner.

Information is included on the subject of painting and decorating, plumbing and pipe fitting, concrete and masonry, carpentry, roofing, glazing and calking, sheet metal, electricity, air conditioning and refrigeration, insect and rodent control, heating, maintenance management, and custodial practices. Even though many different subjects are discussed, each is important in the maintenance of any building—large or small.

This book presents, in concentrated form, information and data gained from the experience of maintenance personnel from many industries. Numerous people were connected, in one way or another, in its preparation. Credit and thanks are due the personnel of the firms contacted for their cooperation in supplying information and illustrations. A sincere effort has been made to give proper credit for this material; however, if any company has been overlooked, acknowledgment is hereby given.

**About the Author . . . .**

Mr. Oravetz is a Registered Professional Engineer in Iowa, Kansas, Colorado, Wisconsin, and Michigan. In addition, he is a licensed Stationary Engineer. For the past 15 years he has been chief of a major engineering and maintenance division with the responsibility of a five-state area. He is also a consultant for the maintenance of buildings and grounds for the Recreation and Parks Department of a major midwestern city.

# Contents

# CHAPTER 1

# Painting and Decorating

Finishes and paint coats wear out and, over a period of time, require replacement. The renewal cycle is not the same in all geographical areas, as interior and exterior painting requirements may vary in accordance with climate, location, and personal preference.

There are many reasons for painting surfaces but the two most important are to improve and maintain appearances. Some surfaces, such as metals, must be painted to prevent rust, corrosion, and deterioration. Painting should produce attractive and long-lasting results. Wood and metal surfaces must be protected from the elements to prevent deterioration and excessive building maintenance costs.

Labor is the major part of a painting job and no economy is redeemed by applying inferior materials and using poor application methods. The use of proper materials and proper application cannot be too greatly emphasized.

Most paints today are purchased ready-mixed, and are labeled for use outdoor, interior, for wood, concrete, masonry, and other specific purposes. Purchasing ready-mixed paint of a good quality assures uniform color, particularly for interior surfaces where the main consideration is to decorate.

Painting justifies economical maintenance as each paint renewal provides for making needed minor repairs which, if neglected, can result in major repair work to the structure. Painting, in effect, generally stimulates inexpensive preventative maintenance and prolongs the life of the structure.

A good maintenance program should be based on local weather conditions and requirements. Dates when surfaces are painted should be recorded and painting cycles should be established to maintain the required appearance standards. Outdoor painting should be scheduled during warm and/or dry seasons in order to keep personnel busy during the entire work year.

## EXTERIOR PAINTING SCHEDULES

In general practice, it has been determined that exterior surfaces require painting every four to eight years, as follows:

White Paint: About once in four years.
Light tints: Four to five years.
Deep tints: Five to six years.
Dark colors: Five to eight years.

Direct sunlight deteriorates paint rapidly and unshaded areas may require painting more frequently than the preceding schedule indicates. The method of application is also a factor in paint wearability. Wood paints normally should be applied in thin coats. Thick coats become brittle and, in time, must be completely removed before repainting. An exception to this is when exterior latex paints are used. Latex paints have a tendency to spread too thin to offer an adequate protective coating unless care is taken during application.

## PAINT DETERIORATION

### Blistering

Blistering (Fig. 1) is usually caused by moisture drawn from the interior of the painted object to the outside by heat from the

*Courtesy National Paint, Varnish and Lacquer Association, Inc.*

**Fig. 1. Blistering is a common cause of paint failure.**

sun. The moisture causes the paint to lift from the wood. The paint will begin to peel after the blisters dry. Frequently, the source of trouble is new plaster, leaky joints, humidity from showers and laundries, etc. Some preventive measures are:

1. Calk window frames and the joints between the siding and jambs.
2. Use proper flashings at all intersections.
3. Apply paint barriers on interior walls.
4. Drain exterior ground surfaces close to buildings and calk and paint the siding and joint sections.

11

## Soiling

All paint surfaces normally gather some dirt. Dirty paint may be washed, but should not be repainted until deterioration has progressed further than merely the gathering or adhesion of dirt.

## Alligatoring

Surface cracks (Fig. 2) that look like alligator leather is caused by applying the second paint coat over a prime coat that is not fully dry, or over a paint coat that has too high an oil content. Alligatored surfaces should be removed by sanding or scraping. Modern paint removers can be used and do not require washing the surface with naptha before painting starts—some brands will remove alligatored paint but require some assistance from a hand scraper.

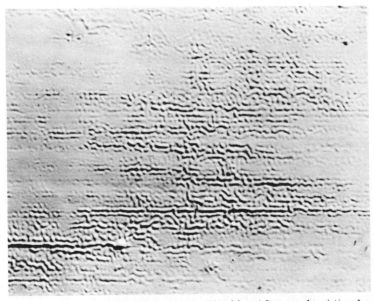

*Courtesy National Paint, Varnish and Lacquer Association, Inc.*

**Fig. 2. Alligatoring is caused by applying paint over a coat that is not completely dry.**

*Courtesy National Paint, Varnish and Lacquer Association, Inc.*

**Fig. 3. Checking is often caused by a paint coat which is too thick.**

## Checking and Crumbling

Temperature changes have a tendency to expand and contract paint. When checking (Fig. 3) appears and cracking does not occur, disintegration takes place by crumbling. When paint is applied in a thick coat and dries out, it loses its elasticity and breaks appear on the surface. Eventually, the edges curl and the paint fails. To prevent checking, *DO NOT* apply thick coats of paint. Prepare the surface by sanding and removing all the old paint before repainting.

## Poor Adhesion (Paint Won't Stick)

It must be remembered that damp wood or masonry will not take paint properly. The weather must be dry and the temperature above 50 degrees. Poor surface preparation results in non-adhesion. Alligatored or peeling surfaces must be removed, and decayed or split boards replaced. All masonry surfaces must be

*13*

brushed. Nails that are protruding should be countersunk and a coating of putty applied. Shellac all apparent knots.

## Chalking

Chalking is a normal process of paint wear, and it causes the finish to rub off to the touch. Chalking does not always offer sufficient reason for repainting exterior surfaces. Rapid chalking occurs when an outdoor finish is applied in damp, cold weather. Before new paint application, the chalk or powdery surface of the paint should be scrubbed with water and a stiff brush. Interior paints rarely, if ever, reach a chalking stage.

## Rapid Deterioration

**Knots**—Knots will sometimes cause blemishes in painted surfaces before the paint wears out. Knots and other bleeding defects should have a coat of shellac or of manufactured knot sealer applied after preparation by sanding to seal the bleed before painting.

**Inferior Paints**—Poor quality paints can deteriorate easily. Therefore, the use of good paint from a reliable manufacturer assures longer life of the painted surface.

**Crawling**—When paint is applied and tries to form into drops (as water does), the term generally applied is *crawling*. This usually occurs when the old paint coat was not properly cleaned of oil and grease. Carefully clean all surfaces before painting.

**Uneven Color**—When paint is spread too thin or applied too lightly, much of its base oil is absorbed and drying occurs with a faded color and poor gloss sheen. Weather conditions such as fog, dew, rain, or frost may also result in the appearance of faded colors. The color defects are not considered serious, and as the paint ages, the defects will usually disappear.

**Mildew**—In warm climates, and in excessively damp locations, mildew (Fig. 4) may occur on painted surfaces. This results in discoloration. Mildewed paint must be scrubbed clean before repainting or the infection may show through the new paint. To remove mildew, wash the surfaces with strong soap and warm water, or with an alkaline cleaner. Then rinse with water and allow to dry before painting. Use rubber gloves when washing with strong soap or trisodium phosphate solutions. If mildew is continuously recurring, it is recommended that paint with a suitable fungicide incorporated into it by the manufacturer be used.

*Courtesy National Paint, Varnish and Lacquer Association, Inc.*

**Fig. 4. Warm, damp locations may cause paints to mildew.**

## PREPARATION OF WOOD SURFACES FOR PAINTING

It must be remembered that even the best of painters cannot do a good job on a surface which is chalky, scaly, alligatored, or otherwise in poor condition. A large percentage of painting failures can be traced to lack of preparation. Just how much preparation of the surface is necessary depends to a degree on the judgment of the person doing the painting. A key to this decision is the basic fact that paint will not cover irregularities. Paint will

15

not hide a scaly area nor will the finished product be any smoother than the surface underneath it.

## Moisture Content

Moisture content in wood should be as low as possible immediately prior to painting. A moisture content of 9 to 14 per cent is preferred for exterior wood surfaces, 5 to 10 per cent for interior wood surfaces, and 6 to 9 per cent for flooring. If exterior wood surfaces are fairly dry but have been temporarily made wet by rain or dew, the painting may start as soon as the surface again becomes dry to the touch.

## Exterior Wood Surfaces

New exterior wood surfaces should be cleaned if there is oil, dirt, asphalt, crayon markings, etc., that may be detrimental to paint adhesion. Normally, new wood is in a condition ready for painting with some exceptions as follows:

**Resin or Pitch Marks**—Remove with scraper or sandpaper. If too soft for a scraper, use turpentine or mineral spirits.

**Loose Splinters**—In some low-grade lumber, splinters appear and these must be removed prior to painting. Some sanding may be required if the splinters leave exceptionally rough edges and the requirements are for smooth appearing work.

**Knots**—A manufactured knot sealer is recommended for sealing knots, although shellac has been used successfully for many years.

**Nail Holes**—Nail holes and imperfections should be filled with putty, but only after the prime coat has been applied and has dried.

**Preparing Painted Surfaces for Repainting**—Blistered, loose, or curled paint should be removed with a wire brush, sandpaper (grit #4 or #4-1/2 heavy), or by a scraper. If chalking only is apparent, and no loose, blistered, or paint curling is present, the surface should be cleaned by using a dry brush or painter's duster.

**Complete Paint Removal for Repainting**—Complete paint removal is not generally necessary unless the paint is so deteriorated that no other procedure is possible. The general method for removing old paint is by scraping and sandpapering. When the paint cannot be removed by sanding or scraping, it must be softened by the use of a paint remover or a painter's blowtorch. *There have been cases where buildings have been set on fire and destroyed by using blowtorches for paint removal, so extreme care must be used when using a "burning" process for removing paint.* The new electric torches available for this purpose are less hazardous and will not char the wood as easily as a blowtorch. *Blowtorches are NOT recommended for paint removal unless extreme precautions are taken against the possibility of fire.*

If a paint remover is used, wash the surface with a paint thinner or a strong solution of soap and water before applying the paint. Some paint removers have a wax base which must be removed because paint does not adhere well to a wax-bearing surface. In addition, some paint removers are toxic and contain inflammable materials. This type should be used in well-ventilated areas only. Nonflammable removers are available where their need is imperative.

## Interior Wood Surfaces

All new interior wood surfaces to be painted should be free of loose dirt, grease, or oil, and all rough spots sanded with fine sandpaper (maximum sandpaper grit to be #2). Imperfections and nail holes should be puttied after the prime coat of paint has been applied. Knots on new surfaces should be thoroughly cleaned, scraped, and sealed with a commercial knot sealer or shellac.

Interior painted surfaces normally can be prepared for painting by washing with a commercial cleaner (detergent) and water, or with plain soap and water. The previously painted surfaces usually remain intact, although they do become dirty and dull. Soil resulting from grease, particularly in kitchen areas, may reveal damaged spots after being washed away. These areas should be spot primed with the type of paint to be used for the finish coat and then lightly sanded after the primer has dried.

Previously varnished woodwork which is to be painted or enameled should receive the following treatment:

1. Apply commercial varnish remover.

2. Scrape or remove all old varnish.
3. Apply paint thinner (turpentine or mineral spirits) to remove any wax that may have been left by the paint remover.
4. After thoroughly drying and a light sanding, apply the paint or enamel.

The application of paint or enamel over varnish may result in excessive alligatoring. Alligatoring may start as checking or cracking, but the break tends to grow wider at the bottom as well as at the top. The top coating contracts, thus exposing portions of the varnish underneath.

## PREPARING METAL SURFACES FOR PAINTING

In order to receive maximum durability from the painted surface of iron and steel, all soil, rust, dirt, oil, grease, and other foreign materials should be removed prior to painting. On those surfaces which are exposed only to normal atmospheric conditions, power wire brushing, scraping, and general sanding are sufficient. These methods are not perfect but, from an economy viewpoint, meet the preparation requirement. The application of a primer with a large proportion of free linseed oil or with a fish-oil base usually provides for a good paint job on steel and iron. Primers with such a base wet the surface and penetrate to the base metal through any residual and adhered rust and/or mill scale. Where metal is to be continuously immersed in water, sandblasting is recommended prior to painting.

The most common methods used in the preparation of metal surfaces for painting are wire brushing, sanding, chipping, scraping, grit or shotblasting, and pickling.

### Chipping, Scraping, and Wire Brushing

Wire brushing on large jobs is usually done with power wire brushes of the portable type, either air or electric driven. A cup-type brush with bristles made of 0.01″ to 0.03″ (diameter) steel wire is recommended.

Experience has shown that applying a power wire brush too long to a given area will not clean any better, but will only polish the steel which does not help paint adhesion. Chipping is time consuming and should be done with care as careless use of scaling and chipping tools leave sharp ridges in the metal which later become focal points for rust spots.

For most steel or iron painting, wire brushing, chipping, and scraping are sufficient if a primer is used that has good wetting properties. Those primers which have a base of free linseed oil or, in some cases, fish oils meet these requirements. Care should be used when cleaning steel welds so as to remove all slag, flux, and wild spatter.

## Preparing Small Metal Areas

When preparing small areas of steel or iron for painting, care must be taken to remove all rust, loose paint, and dirt with sandpaper or steel wool until sound metal is reached. Washing or cleaning with a solvent, such as turpentine, is necessary if grease or oil is on the surface to be painted. Also, a wire brush or steel wool soaked in turpentine will remove most rust spots.

## Preparing New Galvanized Metal for Painting

Galvanized metals should be left unpainted during their early life. However, if painting must be accomplished for the purpose of appearance, a good method to remove any excess grease film is to wash the metal with a solution of household vinegar and water, with a trial mix to determine the proper strength. Wipe dry after cleaning. A major problem in painting new galvanized surfaces is in getting good paint adhesion.

The best method to use for galvanized iron is to allow it to weather for several years, and then clean the surface. To remove grease, dirt, and foreign material after a period of weathering, use turpentine or mineral spirits.

Paint usually fails on galvanized surfaces by peeling or flaking. Therefore, do not build up thick coats of paint, but use only the minimum amount to provide the required appearance. For priming, use a zinc-oxide base paint, or one specifically recommended by a reputable paint dealer. After cleaning, spot-prime any rusted areas and brush the paint out as well as possible. The prime coat should be allowed to dry at least 48 hours before applying the finish coat. A priming coat and a finish coat are usually sufficient for galvanized-iron surfaces.

### Preparing Existing Painted Metal Surfaces

Remove all loose paint and rust spots by using a wire brush and sandpaper. Clean the surface with turpentine or mineral spirits. After the surface has dried completely, use a primer and finish coat, but do not build up excessive layers on well-painted areas.

## PREPARING PLASTER AND WALLBOARD FOR PAINTING

Plastering an entire wall or area requires considerable skill, and a regular plasterer should be called for that purpose. However, plaster patching can normally be accomplished by most persons. There are many good patching plaster mixes that can be purchased, and all can be mixed by adding water to form a workable mix. Some brands set faster than others. Instructions on the container should be read and followed closely.

**Pinholes or Surface Cracks**—Dampen holes or cracks with clean water and press the mixed plaster into place with a putty knife or similar tool, or with your fingers. Smooth out the patched area with a wet sponge or cloth.

**Large Holes or Cracks**—Remove all pieces of old plaster that tend to come loose with some prying. In most instances, this will enlarge the area but will prevent the patch coming loose at a future date. After this is accomplished, undercut the edges around the hole to be patched. Make sure the outer edges of the hole slant inward.

Mix the patching plaster to a workable consistency somewhat like putty and, using a putty knife, work the mixture into the opening. When the patching plaster is near the top of the opening, stop the patching and let it set for several hours, after which time the remaining space should be wetted down with a sponge or cloth dipped in clean water. Complete the patching, using a flat tool (putty knife or trowel) to level the surface.

After all of the area has set and become hard, use a very fine sandpaper to smooth down any uneven sections. After cleaning off any dust or grit, use a glue size or shellac to seal the area. If the sealing is not done, the outline of the patched area may show through after being painted.

Patching plaster sets up in from 10 to 30 minutes (depending on the brand), so do not mix the plaster before all other work has been prepared and you are ready to apply it. Unmixed patching compound will keep indefinitely if kept dry, so be sure the container is sealed against moisture.

## Painting New Plastered Surfaces

New plastered walls or ceilings (excluding patches) should be allowed to dry thoroughly for at least 30 days before painting. Ventilation while drying is very essential, and in cold, damp weather, rooms should be heated. Damaged areas should be repaired as noted for patch plastering.

When new plaster surfaces have thoroughly dried, a coating of primer-sealer should be applied. Commercial synthetic-resin water-emulsion paints (latex or vinyl) serve well as a base, even if an oil-paint finish is planned. These water-mixed emulsion paints provide a good seal. In order to seal the pores in plaster, a sizing varnish, weak glue water, or shellac are used on unpainted plaster areas when a commercial sizing or paint solution is not readily available.

Oil paint of the same color as the final coat is often used for sizing but, if used, it should be diluted with approximately one pint of raw linseed oil to each gallon of paint. The second coat of paint is diluted with a half-pint of thinner (turpentine or other solvent) to each gallon. The finish coat is applied as the paint comes from the container after mixing.

On previously painted plaster surfaces, oil paint is applied directly from the can, after mixing, without dilution by solvents.

## Wallboard

Plasterboard, gypsum-board, and cement-asbestos board are often referred to as wallboard. In order to prepare these surfaces for painting, all joints and nail heads should be filled with a putty preparation usually made and recommended by the manufacturer of the board. In general commercial painting, the plasterboard joints are taped with a special paper seal. The surface is painted after the joints dry and are sanded.

## Repainting

When plaster or wallboard surfaces are in fair condition, preparation is usually limited to brushing with turpentine or mineral spirits, cleaning only the very dirty spots. For greasy walls, use a preparation of warm water into which a small amount of detergent (washing powder or soap) has been mixed.

When repainting, the condition of the plaster or wallboard surfaces determine the number of coats of paint required. Paint failure results from repeated painting, as the paint film increases in thickness. If the surfaces are properly cleaned and prepared, one coat of paint is all that is usually necessary.

## PAINTING CONCRETE AND MASONRY SURFACES

### Exterior Masonry Surfaces

Exterior masonry surfaces are normally painted to keep water out. If the surfaces are comparatively new, have weathered a short time, and are unpainted, scrubbing with clean water may be sufficient for the initial preparation if no defects are present.

### Efflorescence

The white crystalline deposits, called *efflorescence,* on brick walls may be removed with a wire or stiff-fiber brush. If this does not work well, a 10-per cent muriatic or builder's acid water mix should be applied and allowed to remain on the surface for about four minutes and then scoured with a stiff-bristle brush. After application and scrubbing, wash clean with a solution consisting of one-half pint of ammonia to one gallon of water.

When using a muriatic-acid solution, *do not pour the water into the acid*, but pour the acid into the water and mix well. Use old clothes and gloves. The use of goggles during application is advisable.

### Removing Oil and Grease

Oil and grease can be removed by using solvents (turpentine or mineral spirits), steel brushes, steel wool, abrasive stones, or light sandblasting. The reduction of surface smoothness, particularly concrete, is necessary for paint adherence. Removal of concrete glaze can be accomplished in the same way that efflorescence is removed from masonry.

## Joint Repair

Cracks and faulty mortar joints in concrete and stucco must be repaired before painting. For large cracks, prepare an inverted V-shaped cut and fill with mortar. A mix of three parts of mortar sand and one part Portland cement mixed with water to form a putty, makes a good patching compound. Apply the compound into the V-cuts or joints and allow to set, wetting the patches periodically for 48 hours. If using an oil-base paint, allow the patch to cure for a period of about 50 days before paint application.

## Surface Preparation

If a paint similar to the old paint is to be applied, clean off all dirt and dust by brushing, and all loose scale by sanding or wire brushing. If a dissimilar paint is to be applied, such as cement-water paint over an oil-base paint, then all old paint must be removed, using a paint and varnish remover for oil paint and a 10-per cent muriatic-acid solution to remove cement-water paint.

## Concrete Stucco

Portland-cement stucco must be dry before painting and should not be painted until morning dew or dampness has a chance to dry. If concrete stucco has not weathered for at least 10 months, the lime in it can be neutralized by applying a solution of 2-1/2 pounds of zinc sulphate in a gallon of water. The solution is applied to the surface and allowed to dry for several days. Any resultant crystals which may appear on the surface after drying should be brushed away prior to the application of any paint.

## Concrete Blocks

Concrete blocks are considered to be open-textured, and it is therefore difficult to apply conventional varnish or oil paints successfully. These paints *do not* fill the many surface voids. For this reason, latex or cement-water paints are recommended. Latex paint can be worked into the surface with a stiff brush, using a scrubbing action during painting. Latex paints have good alkali resistance and require less aging (normally about 3 weeks) than do oil-base paints.

If a specific requirement exists to paint concrete-block surfaces with an oil-base paint, a preliminary base painting with latex or cement-water paint is recommended in order to fill the voids in the surfaces. When the painting or sealing of concrete blocks is done for the purpose of waterproofing, the side facing the water source should be painted first. Sealing material must be applied in sufficient quantity to fill the porous voids in the block. A stiff-bristled scrub brush will usually work well for this purpose.

If a cement-water paint is used, every effort should be made to secure a consistency similar to rich cream. During painting, add water to maintain the desired thickness. Where waterproofing is desired, the paint must be scrubbed into the surface thoroughly. In two-coat work, a 24-hour drying period between coats is desirable. If the weather is clear and warm, work in the shaded areas when possible. If the surfaces quickly absorb moisture, dampening processes (wetting with water) should be accomplished about an hour before painting. It is recommended that brushes used for the application of water-emulsion paints be soaked in water for at least one hour before use.

## PAINTING METAL SURFACES

### Aluminum and Aluminum Alloy

Clean the metal surface with mineral spirits or turpentine to remove all traces of oil and grease, and then wash with a vinyl-type wash coat, normally a mix of one part acid component to four parts of resin component. This solution can be obtained from most paint dealers. An aluminum-base paint or other paint that is recommended for this purpose by the manufacturer should be used.

### Copper Surfaces

Surfaces should be cleaned with a solvent and then wash coated as indicated for aluminum. For good adhesion, roughen the surface with fine sandpaper and clean off any dust or sand. Use a paint recommended by your paint dealer.

## PAINTING MISCELLANEOUS SURFACES

### Rough Wood and Shingles

Rough wood surfaces normally absorb an excessive amount of paint and, therefore, painting is not generally recommended due to the high cost. However, rough surfaces take shingle stain very well, particularly the dark colors, and two-coat work will normally last six to seven years. Normal coverage of the stain is about 140 square feet to the gallon, depending on the surface to be stained.

### Fabric Covering

Fabric covering over insulation on pipes, ducts, or other equipment, should be given a heavy coat of size to seal the surface. After complete sealing, any desired type of paint may be applied.

### Wallpaper

Wallpaper should be in excellent condition before painting over it. Check for loose or peeling areas. If the paper is not in good condition, it is best to remove it by soaking with warm water and then scraping it off. Before painting, test a small out-of-the-way area with paint to be sure the wallpaper color won't bleed through. If bleeding does occur, it is best to remove the paper. Paint dealers often have wallpaper steamers for rent and, if used properly, the paper should be easily removed. After paper removal, wash the walls free of any remaining paste and size before painting.

## FLOOR FINISHES

Floor finishes are subject to a great amount of abrasion because of traffic, and therefore require more care and must be renewed more frequently than wall areas. A good program of floor care and maintenance must be maintained to assure the maximum life of the floor finish.

### Concrete Floors (Exterior)

It is not recommended that concrete floors be painted before aging at least one year. If it is necessary to paint before this time, remove the alkalinity by washing the surface with a 10- or 15-per cent solution of muriatic acid and water. Prior to painting, rinse the floor with clean water to remove the acid and lime, and allow at least two days for complete drying. Apply a good cement primer, allow to dry 24 hours, and then apply the second coat, brushing well during each application.

## Concrete Floors (Interior, Unpainted)

**Applying Varnish-Base Paints and Enamels**—Prepare the surface by scrubbing well with a mild soap and stiff-bristled brush. Rinse with clear water and allow to dry for at least 24 hours before applying paint. Apply the primer by vigorous brushing and allow to dry for at least three days. Apply undiluted paint as a second coat, and after it has dried sufficiently, apply several coats of liquid floor wax.

**Applying Rubber-Base Concrete Paint**—After the floor is cleaned, etch the floor to receive paint by washing with a solution of one part muriatic acid to three parts water, applying evenly with a broom or long-handled brush. (Use goggles and rubber gloves when applying this solution.) When the floor surface is dry, apply the first coat of rubber-base paint thinned with 20 per cent of mineral spirits to allow good penetration. The second and, if necessary, third coats should be applied at 24-hour intervals. Always have the room where the paint is being applied thoroughly ventilated. Light foot traffic may be permitted after 16 hours, but heavy traffic should be avoided for a period of at least four days.

## Interior Wood Floors

Prior to painting, prepare the surface by brushing and cleaning to remove all dirt and dust, and smooth uneven or rough spots by sanding. Apply a prime coat, allow to dry for 24 hours, and fill nail holes with filler putty. Apply the second coat, brushing the paint along the wood grain. Allow to dry for at least 24 hours, and apply two coats of liquid wax.

If a prime paint is not available, regular floor paint can be used by thinning with 1 quart of spar varnish and 1/2 pint of mineral spirits for each gallon of paint. Most paints dry slowly at low temperatures; therefore, do not paint when the temperature is expected to fall below 40°F.

For severe paint failures where blistering, alligatoring, and flaking occur, strip the old paint coating off the wood with paint and varnish remover. If the paint is only worn, particularly in heavy traffic areas, but is in generally good condition otherwise, showing no flaking or alligatoring, spot prime the worn areas with the regular floor paint which is to be used.

After drying, apply a single coat of paint over the complete floor. After the paint is dry, apply a single coat of liquid floor wax.

### Exterior Wood Floors

Treatment in general is similar to interior wood-floor preparation and painting except that exterior paints are used and the surfaces are not usually waxed.

## VARNISHES AND PAINTS

Paints can generally be placed in the following type categories: water paints, water emulsions, oil paints, enamels, lacquers, cement paints, varnishes, and today's newest type, epoxy paints.

For interior finishes in homes, there are generally three degrees of sheen—*flat, semi-gloss* or *satin finish,* and *high-gloss*. For maximum beauty, the flat finish is preferred for walls and ceilings in living rooms, bedrooms, libraries or dens, and halls. Semi-gloss or satin finish is excellent for kitchens and bathrooms as well as for woodwork, cabinets, and trim. The high-gloss finish provides a hard, lustrous surface where moisture resistance and washability are required, such as in kitchens, bathrooms, laundry or utility rooms, or on woodwork, doors, and cupboards.

It must be noted that interior paints do not wear away as rapidly as outdoor paints unless washed often with water and strong detergents. Interior painting should be kept at a minimum number of coats, and the paint should be brushed out as much as possible to avoid thick coats.

### Water Paints

Water paints are normally those that come as a powder to be mixed with water, but many can be purchased already mixed.

Water paints include *calcimine, casein,* and *texture. Whitewash* is also considered a water paint.

## Water-Emulsion Paints

Latex paint is the most common and best known of the water-emulsion paints. Its use is widespread and effective due to its ability to cover evenly and easily when used by a nonprofessional painter. Painting tools, including brushes and rollers, can be easily cleaned with plain water after being used with this type of paint.

## Oil Paints

These paints have a lead, titanium, or zinc pigment base combined with linseed oil and a drying agent. Thinning agents generally used are either turpentine or mineral spirits. Exterior oil paint normally dries with a gloss and the interior flat paint without gloss. On the flat paints, the liquid is almost all mineral spirits or turpentine, with little or no linseed oil.

## Varnish

Varnishes normally have a sheen or high gloss, but flat varnishes may be purchased. In preparation, a resin is used instead of oil as the vehicle, and when pigments are added, the varnish becomes enamel.

Shellac is a spirit varnish with the thinner or solvent being alcohol. Lacquer, using special solvents, is a special resin varnish, and is highly weather resistant. Most metal paints, such as aluminum, have a varnish base.

## Enamels

Enamels do not have a linseed-oil vehicle, but are made with a varnish base. They dry to a hard, glossy finish that resists abuse and hard washings.

## Masonry Paints

Most common of the masonry paints is the cement type made with Portland cement. It is used for waterproofing and coating porous or cracked masonry surfaces and, if used according to recommended instructions, is highly satisfactory. There are many good masonry paints and sealers on the market today. In use are the waterproofing sealers with silicones, using a wax and resin or varnish base. These are used for exterior surfaces. Rubber paints, made of chlorinated rubber, are in use for interior concrete floors and exterior masonry. This type of paint is highly resistant to excessive surface alkalinity, will not peel readily, and provides a tough, long-wearing surface.

### Epoxy Resin Paints

These paints are relatively new, but have shown excellent chemical resistance. The sole binding agent is resin, with only very small amounts of drying oils or other film-forming materials being used. These paints have excellent abrasion resistance, durability, and hardness.

## PAINT THINNING

Most paint manufacturers provide paint so that it can be applied as it comes from the container, particularly during good weather. If the paint must be thinned, a good grade of mineral spirits or turpentine can be used for almost all paints which are used for interior or exterior work.

When paint thickens in cold weather, a maximum of 1 pint of thinner per gallon is allowable. This practice also can be used when a thin paint coat is required on previously painted surfaces. In three-coat work, 1-1/2 pints of thinner may be added to a gallon of primer if a good brush coat is required.

## SPRAY GUNS

The use of a spray gun to apply paint requires a certain amount of experience which can be gained only through actual operation. There are a number of types of spray guns available, but they fit into two main categories—*pressure feed* and *suction feed*. Generally, the suction-feed type is less expensive, easier to clean, and easier to use with light finish materials. The pressure-feed gun, which works better on heavier materials, is used on big jobs, but this type is higher priced.

When a spray gun is purchased, a determination must be made as to its intended use. A reliable paint dealer will offer suggestions concerning the most practical type to purchase.

Two important factors that determine the success of paint spraying are—(1) the distance the gun is held from the work, and (2) the moving of the spray gun in a straight line. The exact distance the gun should be held from the work is determined by the specific paint job, but the average distance is from 6 to 12 inches. After practice, distances can be worked out for the different types of finishes, such as paint, shellac, varnish, and other materials.

When using a spray gun, it must be remembered that a flexible wrist gives better stroking and a more evenly finished surface. Always press the trigger a moment before the gun moves across the work and release it just before the end of the work, continuing the stroke beyond the point of paint finish. Each stroke should be overlapped. If a large level surface requires spraying, start at the near side and work to the far side so that any overspray falls on the uncoated surface.

When spraying indoors, be sure that there is no open flame and that all materials and furniture that can be harmed by the spray mist are covered. Provide adequate ventilation and use any of the inexpensive types of face masks to prevent excess inhalation of the spray mist.

Outdoor spraying should not be practiced during high winds unless the spraying work is at a great distance from other buildings or parked vehicles. Excessive damage has resulted from windblown spray mist on some jobs and lawsuits for damages have resulted in losses to the painters.

All spray guns must be cleaned immediately after spraying to prevent clogging of the gun. Follow the manufacturer's recommendations. Spray guns can be cleaned with turpentine after using paints and varnishes, denatured alcohol after using shellac, and a lacquer thinner after using lacquer. After using a water- or rubber-base paint, plain water will normally clean the spray gun. Spraying works well when painting screens and reduces labor.

## GLASS-JAR SPRAYERS

Most paints can be used with a glass-jar sprayer if thinned a little so as to provide proper flow through the nozzle. This type of sprayer cannot do the complete job that a pressure-feed sprayer can, but for most small workshop purposes it can accomplish good results with proper use.

New paints can be thinned, but if the paint is old or contains any undissolved particles, it should be strained through a piece of cheesecloth or silk before use. Before using the spray gun, it should be tested on a piece of wood, cardboard, or newspaper, and proper adjustments made to the top screw knob to meet the particular job requirement.

Do not move the sprayer in an arc. Instead, move it back and forth, keeping it the same distance from the surface to be painted. Keep your wrist flexible. Keep the sprayer on a straight line about 12 inches from the work. The paint will ripple or run if the gun is held too close to the work. Practice pressing and releasing the trigger over trial material before any actual painting takes place so as to gain experience.

## SIGN PAINTING

Three basic methods are used for making signs—freehand, stencil cutting, or a screen process.

### Freehand

Freehand lettering is generally performed by establishing guidelines, either horizontal, vertical, and/or inclined by using the basic lettering strokes shown in Fig. 5. To properly execute satisfactory letters, a person must practice and learn the sequence and direction of the strokes used to form the letters.

Lettering can also be performed by mechanical lettering sets or templates that can be purchased at nearly any office and school supply store. Some of the stores in shopping centers sell stencil lettering forms that work well in preparing signs.

### Stenciling

Tube colors or colors in oil prepared to a semipaste consistency, using equal parts of turpentine and linseed oil, work well

31

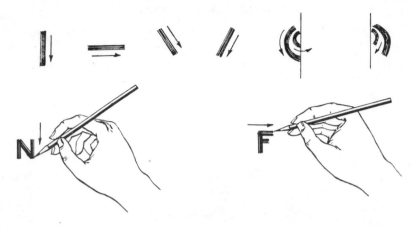

*Fig. 5. Basic lettering strokes.*

for stenciling. However, best results are obtained with regular commercial stenciling ink or paint. When applying stencil color with a spray gun, be sure to mask the adjacent portion of the sign.

When making a stencil layout, it should be from the letters and figures of a master alphabet, working lightly to provide only the letter outlines. Cutting should be performed with a very sharp knife or razor blade, using extreme care to avoid cutting the hand. Use a brush or stencil roller to spread the paint or ink. Most stencil rollers are covered with a short-nap renewable material.

## Screen Process

This type of sign painting is an economical and rapid process when a large number of copies (24 or more) are required. Processing can be accomplished on glass, paper, wood, metal, and cloth. Materials are readily available and complete instructions for silkscreen processes are available from most paint supply stores.

## PAINT STORAGE

It is recommended that large quantities of paint be stored in metal cabinets. A suitable sign, such as *KEEP FIRE AWAY*

—*FLAMMABLE,* should be placed on all cabinets or on doors of paint-storage areas. Paint-storage areas should also be well ventilated to eliminate the possibility of concentration of solvent vapor.

Also to be considered in paint storage is paint identification. Labels or markings must be legible to avoid possible costly mistakes in use.

A common form of stored paint deterioration is the settling of pigments to the bottom of the can. Paint is not damaged if this settlement can be easily re-mixed. Most paints can be stored for at least a year without excessive settlement, except the types having a red-lead base which sometimes settle to a point where they are difficult to mix after a very few months. Some paint users recommend a periodic turning of containers, which does help in preventing pigment settlement.

Stored paint should be used as soon as possible. Storing of paint where temperatures may reach the zero point are, in most cases, not harmful, although the viscosity may be increased. Paint stored in this manner should be left in a well-heated space for a few days before use.

Water-thinned and emulsion-type paints should NOT be stored in freezing temperature locations. Modern resin paints withstand some freezing and thawing cycles, but storage in above freezing areas is recommended.

## SAFETY

Always present in painting work are possible falls from scaffolds and ladders, fires from inflammable volatiles, poisoning from toxic paint materials, and temporary illness from excessive concentrations of vapors from thinners and solvents. It must be remembered that paint solvents have a fairly low flash point. Therefore, all flame should be kept away during application procedures, and paint work areas must be ventilated.

All paint containers should be tightly closed when not being used, particularly at the end of the work day. It is a good practice to remove all paint scrapings and paint-saturated debris from the premises daily.

## Safe Painting Practices

The following painting practices are recommended for safety:

1.  It is recommended that paint-soiled clothing, drop cloths, and cleaning cloths, if still to be used, be placed or stored in well-ventilated steel cabinets.
2.  Smoking should be forbidden in areas where paint is being stored, brushed, or sprayed.
3.  During paint spraying in quantity, a sponge-type respirator should be worn by the painter.
4.  Paint containers must not be heated with any kind of open flame. The containers and contents should be placed in locations heated by steam, hot water, or hot air a few days prior to use. Paint also can be heated by immersing the containers in warm water, the temperature of which should not exceed 145° to 160°F.
5.  If using ladders, be sure they are strong, sturdy, safe, and long enough so you can avoid the dangers of over-reaching. Set ladders firmly at both top and bottom sections. Take time to move ladders along, rather than having to stretch too far and risk a fall. Use a pot hook to hold the paint bucket on the ladder, which permits you to hold onto the ladder with one hand and paint with the other. If planks are used, they must be safely anchored.
6.  Dispose of cotton waste or wiping rags to avoid possible fire by spontaneous combustion. If you must keep them, place them in a covered metal container.
7.  Avoid using painted ladders, as paint often conceals faulty rungs and defective sides. Wooden ladders, particularly if painted, should be inspected very carefully before use.
8.  Do not eat in locations where food may be exposed to lead paint, dust, or fumes, and be sure to wash hands and face carefully before eating.
9.  Avoid frequent use of paint thinners to remove paint from hands or face. Some new commercial hand cleaners contain lanolin and olive oil with a cleaning agent; these cleaners have worked out well for the average person performing painting operations.

## PAINTING HINTS

The following hints may prove useful when painting:

1. On interior rooms, paint the ceiling first, then the walls, and finally the woodwork.
2. Apply paint to the ceiling in areas of about 3 by 4 feet, using a criss-cross action. As you move along, brush from the new area into the previous one.
3. If you are using a brush, select one of good quality for the job. A cheap brush or one that is coarse, fibrous, or dirty, can ruin any paint job.
4. When using a brush, hold it firmly but lightly, and spread the paint over the area with even, moderate pressure, finishing off with an even light pressure. Coat all crevices or irregularities first and brush out oil-type paints completely. When using a paint that is rapid drying, spread it evenly and brush only enough to avoid too much thickness. In corners, protrusions, or rusted areas, use a circular movement followed by cross or parallel brushing. For house painting, a 4-inch brush is the correct size; for trim and corners, use a 1-1/2″ to 2″ pure bristle brush.
5. Prior to use, it is good practice to suspend a new paint brush in raw linseed oil for at least 12 hours, being careful not to bend the bristles. After removal, all the oil should be squeezed out by rubbing the brush against the edge of a smooth board. A turpentine rinse and thorough drying prepares the brush for use.

   When you are through with the brush, clean it with turpentine or mineral spirits if you have been using it for paint, enamel, or varnish. When used for lacquer, clean with lacquer thinner, and when used for shellac, clean with denatured alcohol. When latex or water base paints are used, the cleaning agent should be plain water.

   To store the brush, wrap it in wax paper or aluminum foil, being careful not to bend the bristles. Make every attempt to reuse the same brush for the same type of material—a shellac brush for shellac, a paint brush for paint, and a latex brush for latex paint. A good paint brush will last for years if these procedures are followed.

6. Paint rollers are easy to use. With them you can do a good job on walls and ceilings. Your dealer can supply you with the best type for your purpose, as well as a dip tray for the paint. When painting a wall with a roller, use a brush to start the strip of paint in a corner. Then use the roller, working in strips about 24" wide from ceiling to baseboard.

7. Be sure the paint is well mixed before using. Stir the oil and pigments together thoroughly until the paint is smooth.

8. To estimate the quantity of paint required for interior surfaces, measure the room and add the total length of all four walls. Multiply this figure by the ceiling height for the total area in square feet. Ask your dealer for the coverage figure for the type of paint you select. Divide the total square feet in the room by the coverage per gallon and you will know how many gallons of paint you will need. For ceilings, multiply the length of the room by the width for the square footage. Add this to the total wall area if you are planning to use the same type of paint for the entire room.

Here are some average figures for trim, in square feet, that may be helpful in estimating the amount of paint.

| Window, including casing, one side | 35 sq. ft. |
|---|---|
| Window, sash only, one side | 15 sq. ft. |
| Door, including casing, one side | 35 sq. ft. |
| Door only, one side | 20 sq. ft. |

To compute the number of square feet on an exterior surface, multiply the distance in feet around the building by its average height. Your dealer will tell you or the label on the paint can will show how many square feet a gallon will cover. Divide the total square footage by the coverage per gallon to determine how many gallons will be needed. Generally, allow one gallon of trim paint for each five gallons of body paint.

*Example*: If a building is 20' wide by 40' long, and its average height is 15', then 20' plus 20' plus 40' plus

40′ equals 120′ (total distance around the building) multiplied by 15′ (average height); $20 + 20 + 40 + 40 = 120 \times 15$ = 1800 sq. ft. If the paint covers 400 sq. ft. per gallon, divide the 1800 sq. ft. by 400, and you find that 4-1/2 gallons of paint is required to cover the building with one coat.

9. In warm, dry weather, a 48-hour drying period between coats on exterior work should be sufficient. Interior paints and finishes, where even temperatures are maintained, should be ready for another coat after a 24-hour drying period. Low temperatures or dampness may delay drying periods from 24 to 48 hours.

10. Countersink and putty nailheads before painting exterior work. Such nails may rust out entirely in time, loosening the boards and letting water in. After countersinking, prime the holes with paint and fill them level with a good grade of putty.

11. Remove loose putty on window sash, digging out and replacing those portions that show signs of separating from the glass. This is good insurance against rust on metal sash and against rot on wood sash. Use a prime paint coat on all areas that are to be reputtied, and allow the applied putty to dry several days before final painting.

## FINISHING FIR PLYWOOD

*(Courtesy American Plywood Association)*

It is easy to get professional-looking results on clean, smoothly-sanded plywood when top-quality materials are used and a few simple rules are followed. Follow the manufacturer's directions and prepare the surfaces properly for best results. Carefully clean all surfaces—do not paint over dust, spots of oil, or glue. Fill nail holes and blemishes in exposed edges with spackle or wood paste. Since plywood is generally smooth it's not hard to produce perfect surfaces. Always sand with the grain, using 3/0 sandpaper.

### Interior Finishes

**Paint or Enamel**—Any standard woodwork finish is easy to use if the manufacturer's directions are followed closely. For durability on frequently cleaned surfaces, use washable enamels, following this procedure:

1. After sanding; brush on flat paint, enamel undercoat, or resin sealer. The paint may be thinned slightly to improve its covering power. Fill all surface blemishes with spackle or putty when the first coat is dry. Sand lightly and dust clean.
2. Apply the second coat. For a high-gloss enamel finish, mix equal parts of flat undercoat and high-gloss enamel. Tint the undercoat to the approximate shade of the finish coat. Sand slightly when dry and dust clean.
3. Apply the final coat as it comes from the can.

A two-step finish, without the second undercoat, also may be used.

**Water-Thinned Paints**—Seal the plywood with a clear resin sealer, shellac, or flat white oil paint, to control the grain raise. Paint according to the manufacturer's directions for a sealed surface.

**Stippled Textures**—Textured surfaces are obtained by a heavy face coat of stippling paint after priming. Then texture the paint coat with a stipple brush, roller, or sponge.

**Clear or Colored Lacquer**—Lacquer can be sprayed, brushed, or wiped on. Use the correct type and follow the manufacturer's directions. Sand lightly or steel wool between each coat.

To wipe on brushing lacquer, cover only small areas at a time with a folded pad or soft cloth dipped in three parts of lacquer and one part of lacquer thinner. Rub with a circular motion and carefully blend each patch with the area covered previously.

**Light Stain-Glaze**—A natural finish which mellows the wood's contrasting grain pattern with effective warm colors is always popular. When using any finish which retains the natural grain pattern, carefully select the plywood for pattern and appearance. A four-step procedure is recommended for fine work.

1. *Whiten Panel.* Use pigmented resin sealer or thin interior white undercoat mixed one-to-one with turpentine or thinner. After 10 to 15 minutes (before it becomes "tacky"), dry-brush or wipe with a dry cloth to permit the grain to

show. Sand lightly with fine paper when dry.

2. *Seal Wood.* Apply thinned white shellac or clear resin sealer. Sand lightly with fine sandpaper when dry. Omit the seal coat for greater color penetration in Step 3.

3. *Add Color.* There is no limit to the colors and shades that can be obtained by changing this color coat. Use a tinted interior undercoat, thinned enamel, pigmented resin sealer, or color in oil. With care, light stains might also be used. Apply thinly and wipe or dry-brush to the proper depth of color. Sand lightly with fine paper when dry.

4. *Provide Wearing Surface.* Apply one coat of flat varnish or brushing lacquer. Rub with fine steel wool when dry for additional richness.

**Easy, Economical Finishes**—An easy, inexpensive two-step procedure will result in a pleasant "blond" finish. First, apply an interior white undercoat thinned so the wood pattern shows; tint the undercoat if color is desired. Sand lightly when dry, then apply clear shellac, lacquer, or flat varnish for durability.

The exact, natural appearance of plywood may be retained by applying a first coat of white shellac followed by flat varnish after sanding. Several coats of brushing lacquer may also be used.

Attractive, economical one-coat stain waxes also are available in colors. If dark stain is wanted, first apply a clear resin sealer to subdue the grain.

### Exterior Finishes

Field observations, exposure-fence studies, and weatherometer tests all indicate that the best paint job for regular wood siding also is best for exterior plywood. The high-grade exterior house paints of either TLZ (titanium-lead-zinc) formulation or white lead and oil give excellent service on plywood. The TLZ paints tend to have more lasting appearance qualities. Avoid paints which set to a hard, brittle film.

For complete compatibility between coats, specify prime and finish coats produced by the same manufacturer and formulated as companion products. Allow each coat to dry before the following coat is applied, but painting should be completed as soon as practicable to obtain good adhesion between coats.

**Edge Sealing**—Seal all edges with a heavy application of a high-grade exterior primer, aluminum paint, or a heavy lead and oil paint (100 lbs. white lead paste, 1-3/4 gal. raw linseed oil, and 1 pt. dryer, mixed and applied without thinning). This applies both to exposed edges and edges of panels that are lapped, butted, or covered with moldings.

**Back Priming**—On storage units in unusually damp locations, panels should be back primed with exterior primer.

**Painting Procedure**—The three-coat system is suggested as the best conventional protective coating. A dip or brush application or a top-quality water repellent (toxic or nontoxic) before panels are painted will provide additional protection.

The initial or prime coat is most important. A high-grade exterior primer, thinned with 1 pt. of pure raw linseed oil per gallon of paint (check directions—some paints are not to be thinned with linseed oil) and thoroughly brushed on is recommended. A top-quality exterior aluminum house paint (long-oil spar-varnish type vehicle preferred) makes an excellent outdoor primer for plywood. Conventional paints may be satisfactorily applied over the aluminum prime coat. Greater opacity of the finish coats may be required, however, to completely mask the aluminum primer. Apply at least an initial prime coat as soon as possible. Over this primer, apply the second and third paint coats according to the paint manufacturer's directions.

**Other Finishes**—Top-quality two-coat TLZ paints have been found to perform satisfactorily. However, the same dry film thickness as the three-coat system is necessary.

Stains and natural finishes fail to provide a protective film when applied to plywood; therefore, face checking may be expected. (The permanent waterproof bond between plys, of course, is unaffected.) Natural finishes usually require extra maintenance.

**Marine Uses**—On plywood boats, very satisfactory paint finishes are obtained by using high-grade marine primers, undercoats, and finish coats. Seal the edges and prime the plywood well; for proper adhesion, be sure all paint coats are completely compatible. Finishes which retain some flexibility give the best results. Semi-gloss finishes usually perform better than high-gloss finishes.

CHAPTER 2

# Plumbing and Pipe Fitting

The plumbing maintenance men should have access to the plans of the building or complex of buildings so as to familiarize themselves with the piping and the location of all shutoff valves and water supply piping. The following standards, and those imposed by local plumbing codes, should be observed when working with the plumbing and pipe fittings in any building.

1. All joints must be made permanently gas and watertight.
2. Be careful to provide for contraction and expansion, and install all piping so as to avoid unnecessary strains or stresses on the pipe.
3. Do not pass pipe through structural members of buildings or notch or cut such members unless absolutely necessary. If this must be done, then adequate reinforcement must be provided.
4. Pipe cleanouts should be located at each change of direction and near the bottom of each vertical stack or waste. Provide cleanouts at locations where they can be conveniently reached.
5. Each trap must be protected with a stack vent, back vent, circuit, or loop vent, or other method to prevent trap siphonage. An important safeguard to the health of the people occupying the building is the proper venting of the drainage system.
6. Install vent pipes in accordance with local codes, but never install one that is less than four inches in diameter in cold climates. A vent pipe, if too small, can be closed by frost.

7. Contraction and expansion must be considered when installing pipe under walls. In light construction, wrapping with asphalt-impregnated paper or straight asphaltum should be sufficient, but in heavy masonry construction, a pipe sleeve is recommended.

8. The recommended pipe slope for various sizes of horizontal pipe are as follows (unless local codes dictate otherwise):

    2" and less .... 1/4" per foot
    2" to 6" ...... 1/8" per foot
    over 6" ....... 1/6" per foot.

9. When installing drainage and vent pipes, cast-iron or malleable-iron fittings with screwed pipe are permitted by most local plumbing codes. Cast-iron soil pipe with lead-calked joints is recommended, however.

## PLUMBING AND PIPE-FITTING TOOLS

### Pipe Wrenches

One of the most common plumbing tools is the pipe wrench. These wrenches are used for turning pipe and fittings, and come in various types, sizes, and lengths (Fig. 1). The recommended wrench size for various uses is as follows:

1. A 10-inch wrench for pipe up to 1" in diameter.
2. An 18-inch wrench for pipe from 1" to 2" in diameter.
3. A 24-inch wrench for pipe 2" to 3" in diameter.
4. A chain wrench for pipe 3" and larger.
5. Chain pipe wrench for any pipe, conduit, or irregular shaped fitting.

### Pipe Cutters

Two types of cutters are normally used for pipe under 2-1/2" in size—the *single-wheel cutter* (Fig. 2) used when a full turn can be made on the pipe being cut, and the *three-wheel cutter* used when turning is restricted and difficult. A *geared pipe cutter*

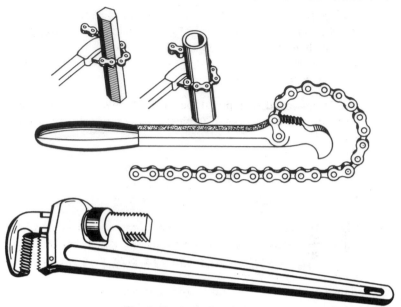

**Fig. 1. Typical pipe wrenches.**

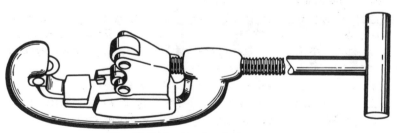

**Fig. 2. A single-wheel pipe cutter.**

(normally electrically operated) is required for pipe over 2-1/2″ in size. The most common cutter is the single-wheel cutter.

## Reamers

When a burr is left on the inside of the pipe by the cutter, a *reamer* (Fig. 3) is used to remove it. Sizes of reamers vary in

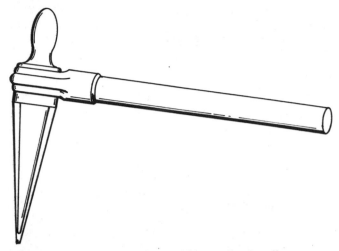

*Fig. 3. A pipe reamer with a ratchet handle.*

accordance with pipe sizes and normally are used either with a
brace (for small pipe) or a ratchet handle (for large pipe).

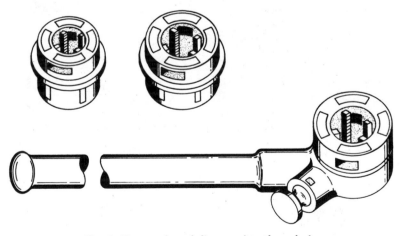

*Fig. 4. Pipe stock and dies used to thread pipe.*

## Stocks and Dies

Pipe threading is done with *stock and dies* (Fig. 4). The pipe
is placed in a vise and the threading accomplished with a plain

or ratchet-type stock-and-die set. When preparing threads on pipe over 2-1/2″ in diameter, a geared (automatic) pipe threader is recommended.

## Vises

*Toothed hinged vises* (Fig. 5) are in most common use and are bench or tripod mounted for shop or on-the-job projects. Inserts of rubber, leather, or lead are recommended when working with plated or polished pipe which can be damaged by the teeth in the jaws.

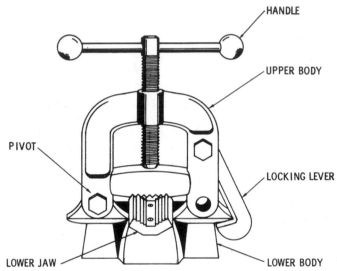

Fig. 5. A hinge-type pipe vise.

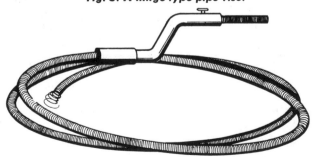

Fig. 6. A typical drain and trap auger.

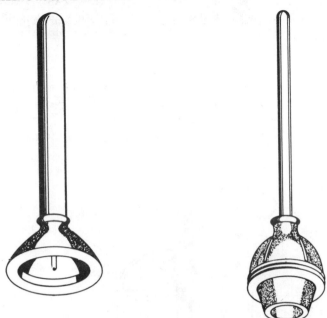

*Fig. 7. Two types of force cups that can be used to clear stoppages in drains and traps.*

### Drain and Trap Augers

This tool (Fig. 6) is made of coiled tempered steel wire and is used to clear obstructions in waste pipes and traps. It is flexible, comes in a variety of lengths and thicknesses, and easily follows piping bends.

### Force Cup (Plumber's Friend)

The *force cup* (Fig. 7) is the simplest of all plumber's tools to use. It is placed over the drain opening and forced up and down. The upward suction or downward pressure may clear the stoppage. The fixture to be cleared should always contain some water.

### Sewer "Snakes"

A *sewer snake* (Fig. 8) is generally used to clear underground sewer lines, including those under buildings. They are normally made of flat spring steel with an average length of 50 feet, although longer lengths can be purchased. Sewer snakes are sometimes used in place of a root-cutting tool to clear outside drains or sewers.

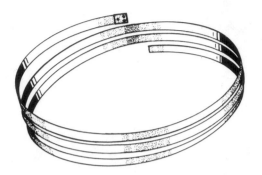

*Fig. 8. A sewer "snake" used to clear sewer lines.*

## Closet Augers

The *closet auger* (Fig. 9), used to remove obstructions in the water closet trap and the bend below, consists of a 30-inch piece of brass tubing curved at the bottom like a cane handle. Inside the tubing is a steel coiled spring about 6-feet long fitted with a crank handle. As the handle is turned, the tempered piece of spring steel works its way down into the toilet bowl, as shown in Fig. 10.

*Fig. 9. A closet auger.*

## Calking Tools

*Packing irons* (Fig. 11) are used to tamp oakum into the hub and spigot of cast-iron soil pipe when making joints. *Calking irons* (Fig. 12) are used to pack lead wool or poured lead against the oakum in the hub and spigot of a calked piece of cast-iron soil

47

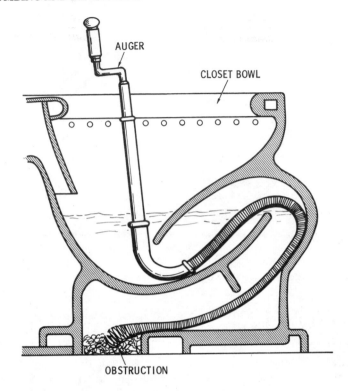

Fig. 10. Details of a stoppage in a closet bowl being cleared with an auger.

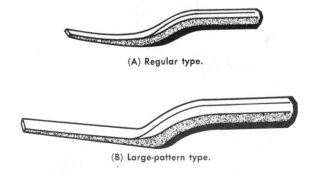

(A) Regular type.

(B) Large-pattern type.

Fig. 11. Packing irons used to tamp oakum into soil-pipe joints.

(A) Outside type.　　　　　　　(B) Inside type.

*Fig. 12. Calking irons.*

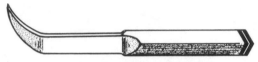

*Fig. 13. Pickout iron.*

pipe. A *pickout iron* (Fig. 13) is used to remove the oakum and lead from a joint that has been made up.

## Tube Cutters

The *tube cutter* (Fig. 14), as the name implies, is used to cut tubing to the required length. Most cutters have a reamer attached, as shown in the sketch, that is used to remove the burr from the inside of the tubing. These cutters can be purchased in several styles and sizes. Hack saws can also be used for cutting tubing, but special care must be taken to make sure the cut is made square.

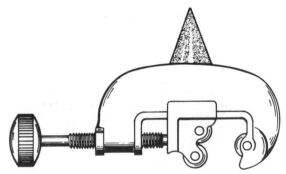

*Fig. 14. A typical copper-tube cutter.*

## Flaring Tools

A *flaring tool* is used when compression-type copper fittings are placed on flexible tubing to match the machined ball of the fitting. This tool generally is a two-piece unit, as shown in Fig. 15.

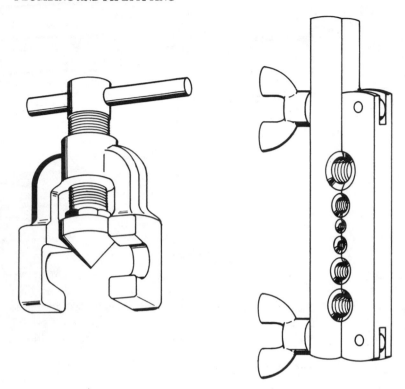

*Fig. 15. A two-piece copper-tube flaring tool.*

## Tube Benders

In order to get a kink-free bend when working with copper tubing, a *tube bender* is used. Some soft-temper copper can be bent without this tool, but benders are recommended. In general use is the lever type, shown in Fig. 16, which can be obtained in several sizes.

## Torches

An *acetylene torch* is generally used by tradesmen for the purpose of heating copper tubing and fittings when soldered joints are specified. However, other types of torches are available, including gasoline, hand-held LP (liquified petroleum), and the small tank type.

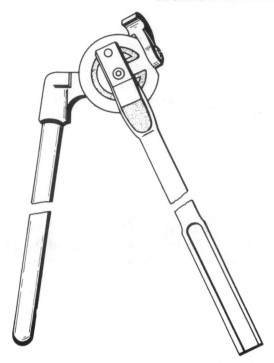

*Fig. 16. A lever-type copper-tube bending tool.*

## Miscellaneous Tools

In addition to those tools generally connected with the plumbing trade, the plumber frequently uses tools common to other tradesmen, such as:

**Steel Measuring Tapes and Folding Rules**—Used to take measurements.

**Chisels**—The cold chisel is used to cut pipe, and the woodworker's chisel to notch wood members to place piping.

**Levels and Plumbs**—Used to determine the grade of pipe lines and to level fixtures.

**Hammers**—The carpenter's hammer is used to drive and pull nails in structural wood members.

**Brace and Bit**—It is often necessary to bore holes in the wooden members of buildings for the passage of tubing and pipes. Most plumbers have a brace and auger bits for this purpose. The brace is also used to hold a reamer to remove burrs from the interior of pipe.

**Portable Electric Drills**—A portable electric drill is used for many different operations and comes with one of 3 different types of chucks. The simplest chuck is that which is tightened by hand. Another type is tightened with a socket or Allen wrench. The third type of chuck is locked with a geared key and is commonly known as the *Jacobs chuck*. Best of all is the geared key type, which is strong, easy to use, and long-lasting. An electric drill is sized by the maximum diameter of the bit, or bit shank, that can be held in the chuck.

Always use a brad awl or nail to mark a point which is to be the center of a hole made in wood. For metal, mark the starting point with a center punch. The point of the bit should be placed in the starting hole before the motor of the drill is started. If the motor is started first, the whirling bit often leaves scratches on the surface of the material before settling in the proper place. When drilling completely through a piece of wood or metal, the stock should be backed up with a block of wood so as to prevent splintering or excessive burrs.

**Saws**—Carpenter's saws are used to cut openings for pipe in wood building sections, and the hack saw to cut copper tubing or iron pipe.

**Wrenches**—A variety of wrench types is used, in addition to pipe wrenches, to turn unions, valves, and plugs, and to hold nuts and bolts. Included are crescent, box, open-end, and the monkey wrench.

## CAST-IRON PIPE

Cast-iron pipe is available in two weights, and is used for drains, sewers, and vent-and-stack systems. *Standard-weight pipe* is adequate for most construction, but the *heavy-weight pipe* should be used under roadways and for tall stacks.

The smallest diameter cast-iron pipe in general use is 2 inches. Commonly used for soil branches and stacks is the 4-inch diameter pipe, with the minimum diameter for certain jobs being 3 inches, although the 3-inch size is not generally recommended.

The average length of cast-iron pipe is 5 feet, and the pipe comes in *single-hub* (bell) with one spigot end (Fig. 17A) and in

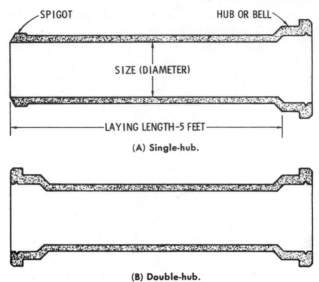

SPIGOT         HUB OR BELL

SIZE (DIAMETER)

LAYING LENGTH–5 FEET

(A) Single-hub.

(B) Double-hub.

*Fig. 17. Cast-iron soil pipe.*

*double-hub* with two hub (bell) ends (Fig. 17B). When using cast-iron pipe, the workman should always have several lengths of double-hub type available. When a requirement exists for a piece of pipe shorter than 5 feet, it is cut from the double-hub section. Economy is practiced by this method, as the cut yields two usable pieces of pipe, each with a hub. Cutting a single-hub pipe may result in the spigot end being wasted.

### Cast-Iron Tees and Bends

When designed for use in a vent line, T's are called *straight T's,* and when used to carry drainage, are called *sanitary T's.* Both types are used for connecting vent lines to stacks and for connecting threaded pipe to branch drains.

**Sanitary T**—A sanitary T (Fig. 18) is commonly used in the main stack for takeoff of a branch drain. The curved section into the T makes a smooth transition of flow direction.

**Straight T**—When the side takeoff and through sections are identical in size, the straight T (Fig. 19A) is described by that size. When the side takeoff is smaller, the fitting is called a reducing T (Fig. 19B) and both dimensions are used, with the straight-through dimension given first.

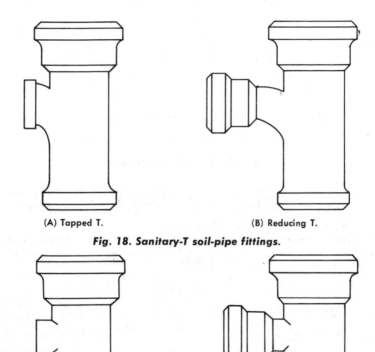

(A) Tapped T.　　　　　　　　　　　(B) Reducing T.

*Fig. 18. Sanitary-T soil-pipe fittings.*

(A) Tapped T.　　　　　　　　　　　(B) Reducing T.

*Fig. 19. Straight-T soil-pipe fittings.*

## Y Branches

This type of fitting is generally used to join two sanitary branches and to connect a single branch to a main. The 45- and 90-degree types are most commonly used, and both permit a smooth flow direction transition.

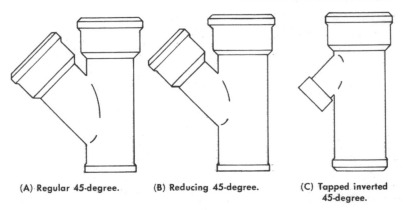

(A) Regular 45-degree.    (B) Reducing 45-degree.    (C) Tapped inverted 45-degree.

**Fig. 20. Y-branch soil-pipe fittings.**

The *45-degree Y-branch* cast-iron fittings (Fig. 20) have the side takeoff at a 45-degree angle and are normally of the regular, reducing, or tapped inverted types.

*90-degree branches* are available in a variety of combinations and patterns, as shown in Fig. 21, which include *regular, reducing, double, box,* and *upright* types.

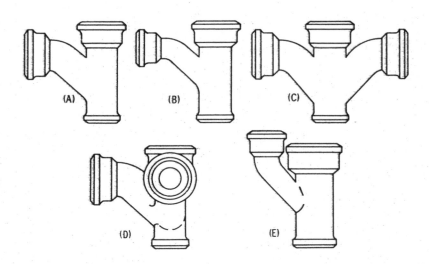

(A)    (B)    (C)

(D)    (E)

**Fig. 21. 90-degree branch soil-pipe fittings; (A) Regular; (B) Reducing; (C) Double; (D) Box; (E) Upright.**

55

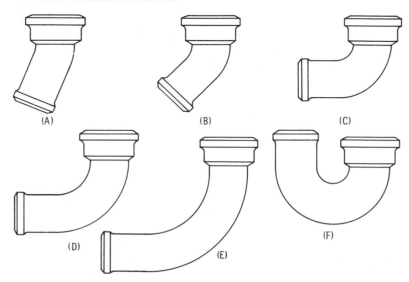

Fig. 22. Soil-pipe bends; (A) 1/16 regular; (B) 1/8 regular; (C) 1/4 regular; (D) 1/4 short sweep; (E) 1/4 long sweep; (F) 1/2 return.

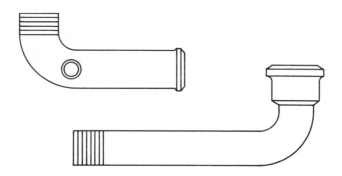

Fig. 23. Soil-pipe closet bends.

*Bends* (Fig. 22) are used to change the direction of the soil-pipe line and are classed as to the degree of turn, such as 1/16, 1/8, 1/6, 1/5, 1/4, or 1/2. The direction of the line change for a 1/16 bend is 22-1/2 degrees, for 1/8 is 45 degrees, for 1/6 is 60 degrees, for 1/5 is 72 degrees, for 1/4 is 90 degrees, and for a 1/2 bend the direction changes by 180 degrees.

56

## Miscellaneous Fittings

**Closet Bends**—*Closet bends* (Fig. 23) are made in different styles to fit specific requirements. They are classed as a special fitting for placing in a soil-line branch and for connecting to a water closet.

**Traps**—*Traps* (Fig. 24) provide a water seal to prevent entrance into the building of sewer gases from soil or waste out-

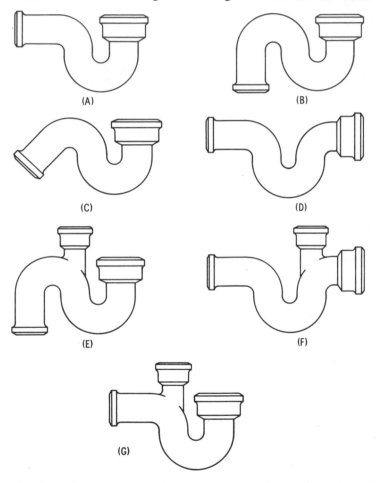

**Fig. 24. Soil-pipe traps: (A) P-type; (B) S-type; (C) 3/4 S-type; (D) Running-type; (E) Vented S-type; (F) Vented running-type; (C) Vented P-type.**

lets. The *P-trap* is most commonly used. Other types include the *S*, 3/4*S*, and the *running trap*.

**Offsets**—*Offsets* (Fig. 25) are used to carry lines past building beams, joists, or other obstructions.

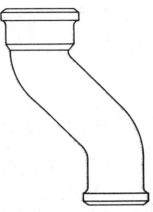

*Fig. 25. An offset soil-pipe fitting.*

**Increaser**—As the name implies, an *increaser* (Fig. 26) is used to increase the size of a pipe line, particularly stacks and vents. This cast-iron fitting is often used at the top of the main vent and stack.

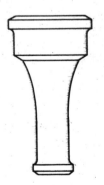

*Fig. 26. A soil-pipe increaser fitting.*

**Clean-out Plugs**—The *clean-out plug* (Fig. 27) is an essential part of any drainage system. This fitting, which has a brass screw plug, is calked into the hub of a cast-iron pipe. The brass plug can be removed when necessary to clear the line of obstructions with a cleaning rod or tool.

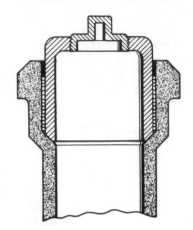

Fig. 27. A soil-pipe clean-out
plug.

## Threaded Pipe Fittings

Fittings of this type are used on either steel or wrought-iron pipe, and normally are made of either cast or malleable iron, or of forged steel.

**Tee's**—*Tee's* (Fig. 28) are used to make a 90-degree branch to the main line.

**Elbows**—*Elbows* (Fig. 29) are used to change directions in pipe lines.

**Unions**—*Unions* (Fig. 30) are used to join two pieces of pipe so they can be easily connected or taken apart.

**Couplings**—A *coupling* (Fig. 31) is a female threaded piece of pipe used for straight pipe connections.

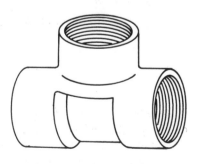

Fig. 28. A cast-iron threaded T.

59

(A) 90-degree elbow.  (B) Reducing elbow.  (C) Street elbow.

**Fig. 29. Cast-iron threaded elbows.**

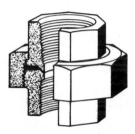

**Fig. 30. A typical union.**

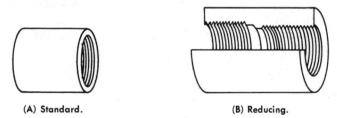

(A) Standard.  (B) Reducing.

**Fig. 31. Couplings.**

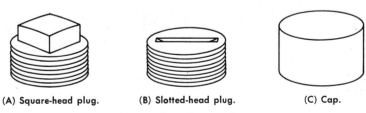

(A) Square-head plug.  (B) Slotted-head plug.  (C) Cap.

**Fig. 32. Pipe plugs and caps.**

**Plugs and Caps**—*Plugs* and *caps* (Fig. 32) are used to close openings in other fittings or pipe.

**Bushings**—*Bushings* (Fig. 33) are used to connect the male section of a pipe to a fitting of a larger size.

(A) Hexagon.  (B) Face.

**Fig. 33. Pipe bushings.**

**Nipples**—*Nipples* (Fig. 34) are used to join fittings or to extend a fitting.

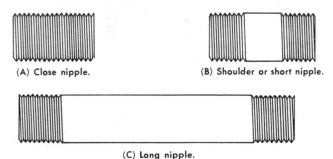

(A) Close nipple.  (B) Shoulder or short nipple.

(C) Long nipple.

**Fig. 34. Pipe nipples.**

## How to Read Reducing Fittings

To assist in "reading" reducing fittings, a variety of types most commonly required for piping systems is illustrated in Fig. 35. In these illustrations, each opening of the fitting is identified with a letter which indicates the sequence to be followed in reading the size of the fitting. In designating the outlines of reducing fittings, the openings should be read in the order indicated by the sequence of the letters *A*, *B*, *C*, and *D*. For reducing fittings with a side outlet, the size of the side outlet is named last.

*Example*—A cross having one end of the run and one outlet reduced would be designated as:

A  B  C  D
2-1/2″ × 1-1/4″ × 2-1/2″ × 1-1/2″

Simply name the largest opening first and then name the other openings in the order indicated.

*Note—Although all but one of the fittings shown in Fig. 35 are the threaded type, the same rules apply to the "reading" of flanged, welded, soldered, and other types of fittings.*

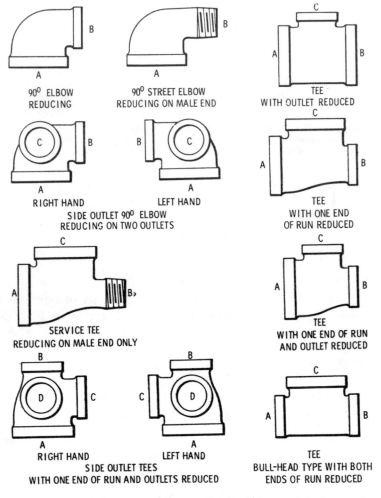

90° ELBOW
REDUCING

90° STREET ELBOW
REDUCING ON MALE END

TEE
WITH OUTLET REDUCED

RIGHT HAND     LEFT HAND
SIDE OUTLET 90° ELBOW
REDUCING ON TWO OUTLETS

TEE
WITH ONE END
OF RUN REDUCED

SERVICE TEE
REDUCING ON MALE END ONLY

TEE
WITH ONE END OF RUN
AND OUTLET REDUCED

RIGHT HAND     LEFT HAND
SIDE OUTLET TEES
WITH ONE END OF RUN AND OUTLETS REDUCED

TEE
BULL-HEAD TYPE WITH BOTH
ENDS OF RUN REDUCED

**Fig. 35. Sequence of designating the openings**

## Right- and Left-Hand Fittings

To clarify the terms *Right Hand* and *Left Hand*, refer to Fig. 36 for the method used to correctly designate these types of fittings. While standing with your back to the stack, the closets to the left of the horizontal drainage line require *left-hand* fittings and those to the right require *right-hand* fittings.

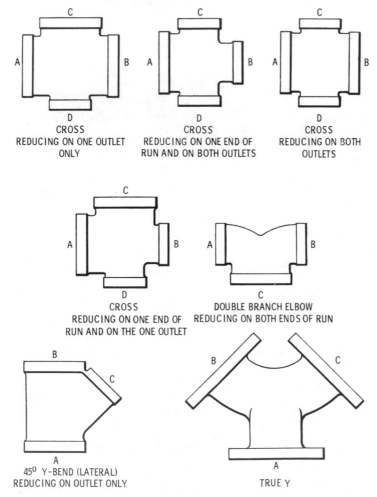

CROSS
REDUCING ON ONE OUTLET
ONLY

CROSS
REDUCING ON ONE END OF
RUN AND ON BOTH OUTLETS

CROSS
REDUCING ON BOTH
OUTLETS

CROSS
REDUCING ON ONE END OF
RUN AND ON THE ONE OUTLET

DOUBLE BRANCH ELBOW
REDUCING ON BOTH ENDS OF RUN

45⁰ Y-BEND (LATERAL)
REDUCING ON OUTLET ONLY

TRUE Y

*on various types of reducing fittings.*

## VITRIFIED CLAY PIPE

Vitrified clay pipe (sometimes called *vitrified tile*) is used for sewer lines and storm drains. The sections come with an asphaltic end seal fitting (called a *slip-seal*) for easy installation, as well as the common hub-end and spigot-end type. Fittings are as varied as those for cast iron, but some types are more difficult to obtain. Pipe diameters range from 4 to 42 inches. As with cast-iron pipe, good workmanship is needed to insure tight seals at the joints.

## PLASTIC PIPE

The use of plastic pipe is becoming common, particularly in locations where highly-corrosive waste must be disposed of. Some industrial plants are using plastic pipe for their water distribution systems, particularly for underground lawn-sprinkling piping. Some newer homes are equipped with plastic pipe and accessories, including P-traps. Plastic pipe is obtainable in *rigid*, *semi-rigid*, and *flexible* types in sizes ranging from 1/2 to 6 inches in diameter. It is best to follow the manufacturer's recommendations when installing plastic piping.

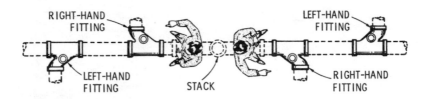

*Fig. 36. Method of determining right- and left-hand fittings.*

## CUTTING AND REAMING PIPE

It has been proven that some stoppages in waste pipe can be traced to poor reaming of the cut pipe. The burrs catch threads, hair, and other wastes, and eventual stoppage is the result.

Before cutting pipe, mark the location to be cut with a file or crayon. Place the pipe in a pipe vise, locking it securely. Close the jaws of the cutter against the pipe, turning the handle clock-wise. Continue turning the handle clockwise as the cut is being made, applying thread-cutting oil to the pipe and cutter wheels. After the pipe is cut, ream it to remove all inside burrs. Remove any burrs on the outside of the pipe by filing to prepare the pipe for the thread-cutting dies. When threading pipe, be sure the dies are not excessively worn or nicked, and be sure to use the right die for the pipe to be threaded. A general rule of thumb when threading pipe is to cut the threads until the cut end of the pipe is about 1/2-inch past the base of the die.

When joining pipe, inspect and clean the threads, using an old finger-nail brush, a small stiff paint brush, or a toothbrush, cleaning both the male and female threads. Pipe or thread "dope" lubricant should always be used (Fig. 37) when making up screwed joints. Apply it to the male end of the joint only, not to the female thread in valves or fittings. The excess will then be forced to the outside and not into the line where it is likely to cause trouble. When joining pipe, it is proper to make several hand turns with

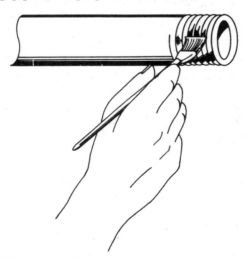

*Fig. 37. Applying pipe-thread lubricant (pipe dope).*

the pipe before using a wrench. Use the wrench to tighten the joint. A rule of thumb dictates that about three threads will show on the pipe after it is tightened if the threading was proper.

## Supports for Threaded-Pipe Installation

All horizontal pipe runs should be supported at 10-foot intervals to prevent strain òn the joints. Horizontal takeoffs from vertical joints should also be supported near the joint. Short vertical runs are not usually supported unless conditions so require.

## Cutting Cast-Iron Soil Pipe

When cutting cast-iron soil pipe, the normal practice for the standard-weight type is to make a cut 1/16-inch deep completely around the pipe with a file, hacksaw, or a cold chisel. The pipe is then struck with a hammer causing it to snap at the score point.

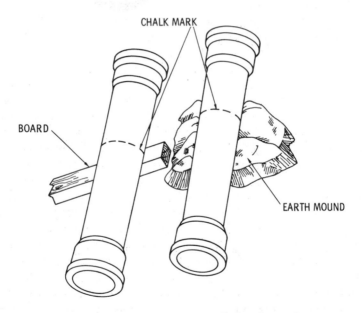

*Fig. 38. Supporting cast-iron soil pipe for cutting.*

For extra-heavy pipe, mark entirely around the pipe with chalk or score it with a file. Lay the point to be cut on a board or mound of earth, as in Fig. 38, so the pipe can be easily rotated. Score with a cold chisel, striking lightly with a hammer until the entire circumference has been scored. Continue this process, striking harder until the pipe breaks at the score mark.

## CALKING CAST-IRON PIPE

Before calking, be sure the spigot and hub ends of the pipe are dry, as moisture has a tendency to make hot lead sputter and fly about. Clean away all foreign materials such as dirt, grease,

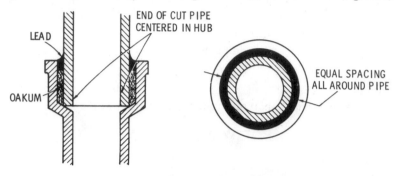

*Fig. 39. Calking cast-iron soil-pipe joints.*

heavy dust, etc. Center the spigot end of the pipe in the hub of the section to which it is to be joined. Carefully pack twisted or spun oakum into the hub, completely around the entire end of the pipe. Continue packing the oakum into the hub in layers until it is about 1 inch from the top, making sure the oakum is tightly packed. This is shown in Fig. 39. *Note: If oakum is not available, cotton braid or jute may be used. Use a packing iron for all packing procedures.* Pour molten lead into the joint with a ladle, making sure the ladle contains enough lead so that the joint can be made in one pouring. Allow about 2 minutes for the lead to harden. Drive the lead down onto the packing and into surface contact with both the spigot and hub end of the joint with first an outside and then an inside calking iron. Do not pound too heavily, as the pipe may crack if the calking is made too tight. This would require replacement of the damaged pipe.

The following are the approximate material requirements for calking soil-pipe joints:

<pre>
3-inch pipe............3 ft. of oakum, 1-3/4 lbs. of lead
4-inch pipe............5 ft. of oakum, 3-1/4 lbs. of lead
6-inch pipe............8 ft. of oakum, 4-3/4 lbs. of lead
</pre>

*Using lead can be dangerous.* One of the main causes of injury when using lead is burns caused by the presence of moisture either on the lead or in the ladle or melting pot. *Be sure the lead, ladle, and melting pot are free of all moisture.* Moisture turns into steam when heated and will cause the lead to be thrown violently from the container. These splashes can result in serious burns.

When the lead melts, impurities will form on top as a slag. This slag must be removed, by scooping off with the ladle, and discarded before pouring a joint.

## HOT WATER HEATERS

The most common hot water heaters are the gas, oil, and electric types. All should be equipped with a pressure relief valve and with a drain at the base for periodically draining the sediment. All repairs should be made in accordance with the manufacturer's instructions.

### Electric Storage Heater

Heating water by electricity normally costs more than with gas or oil, but it is cleaner and is generally safer. In addition, an electric hot water heater may be placed at locations away from the chimney since there are no products of combustion to be eliminated and no smoke or flue pipe is necessary. The electric heater is completely automatic, with the heating elements thermostatically controlled (Fig. 40).

### Gas Hot Water Storage Heater

In a gas storage heater (Fig. 41), a thermostatically controlled gas burner provides the heat. The storage tank is of galvanized iron, may be copper or glass lined, and is enclosed with an insulated jacket. The gas-type storage heater is efficient, comparatively inexpensive, and has a rapid recovery rate as hot water is used. The water temperature is normally maintained from 110 to 180 degrees, depending on the thermostat setting.

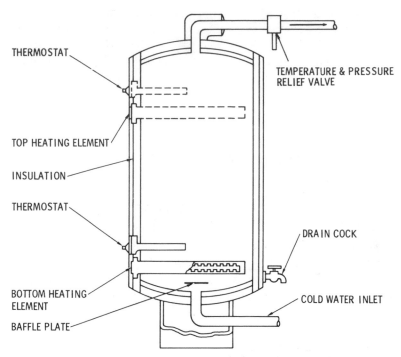

THERMOSTAT

TEMPERATURE & PRESSURE RELIEF VALVE

TOP HEATING ELEMENT

INSULATION

THERMOSTAT

DRAIN COCK

BOTTOM HEATING ELEMENT

COLD WATER INLET

BAFFLE PLATE

*Fig. 40. Cross section of a typical electric hot-water heater.*

## Oil Storage Heater

Oil hot water heaters are used in locations where gas is not available and electric heaters are considered too expensive to operate. It is similar in construction to the gas storage heater, except that it's source of heat is oil. A pressure-type oil burner is generally used in this type of water heater.

## Hot Water Use

Use of water varies in accordance with the building use. In some instances, apartment houses have averaged as high as 120 gallons per person per day, but this usage is above average and not all was hot water. In large office buildings, the average daily consumption of water for all uses can be calculated at 42 gallons per day per occupant.

Rule of thumb calculations indicate the maximum per-occupant use of hot water is about 2 gallons per hour for schools, 8

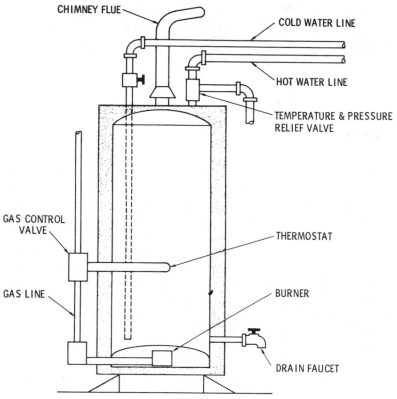

CHIMNEY FLUE

COLD WATER LINE

HOT WATER LINE

TEMPERATURE & PRESSURE RELIEF VALVE

GAS CONTROL VALVE

THERMOSTAT

GAS LINE

BURNER

DRAIN FAUCET

*Fig. 41. Cross section of a typical gas hot-water heater.*

gallons per hour for living quarters, and 4 gallons per hour for offices.

## PLUMBING FIXTURES

Plumbing fixtures generally consist of water closets, water-closet tanks, lavatories, showers and/or shower stalls, urinals, bathtubs, laundry tubs, water fountains, and kitchen sinks. The following pipe sizes are recommended for use on average type installations unless specific local plumbing codes designate otherwise.

Lavatories ........................................................3/8 inch
Urinals (with flush valves) ...........................1/2 inch

Laundry Tubs ................................................1/2 inch
Drinking Fountains ..........................................3/8 inch
Showers ......................................................1/2 inch
Water-Closet Tanks ........................................3/8 inch
Water Closets (with flush valves) ........................1 inch
Kitchen Sinks ..............................................1/2 inch
Commercial-Type Restaurant Scullery Sinks ....1/2 inch

## Lavatories

Lavatories vary in type, size, and color. Offices and residences normally are supplied with vitreous-china types, but those made of enameled cast-iron are often installed in other kinds of buildings. Care should be used in the handling of fixtures as they are easily damaged. Harsh cleaners or abrasives should not be used as they are likely to scratch and mar the surfaces.

Each lavatory must be provided with a P or S trap in the waste line to act as a seal and to prevent waste-gas odors from returning to the building. Most common of the lavatories are the *wall-hung type* with either a pop-up or plug-type drain. The pull-out plug drain (Fig. 42) is less expensive than the pop-up type.

The pop-up drain is equipped with a lever in the waste line connected to a plunger to activate the stopper. The detail sketch in Fig. 43 shows the assembly of the unit.

## Urinals

It is recommended that urinals be installed in areas which have a nonabsorbent floor. If urinals must be installed in rooms with absorbent-type floors, a proper floor covering should be provided. Traps and flushometers are similar to those installed on water closets. Although flushing is possible with the use of an overhead tank, the diaphragm-type flush valve is preferred if quietness is not a factor.

It is recommended that urinals be cleaned daily with a disinfectant or strong soap. Place a screen in the base to catch foreign objects, such as cigar and cigarette butts, chewing gum, cigarette and chewing-gum wrappers, etc. Screens may not intercept all foreign objects, however, and some may slip into the trap. A force cup (plumber's friend) should be used to clear the obstruction.

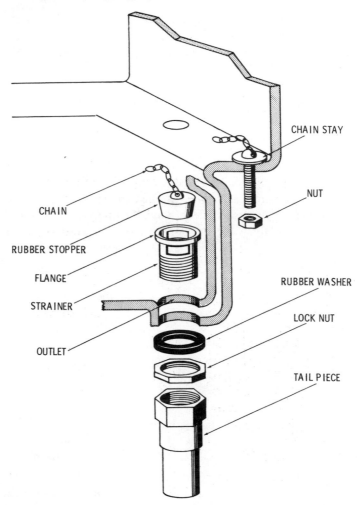

CHAIN STAY

NUT

CHAIN

RUBBER STOPPER

FLANGE

STRAINER

OUTLET

RUBBER WASHER

LOCK NUT

TAIL PIECE

*Fig. 42. Details of a pull-out plug-type drain.*

If this fails, it may be necessary to disconnect the trap and use a "snake" to free the line.

### Diaphragm Flush Valves

A water supply pipe with a diameter of 1 inch or more should be used when installing a diaphragm-type flush valve. This valve

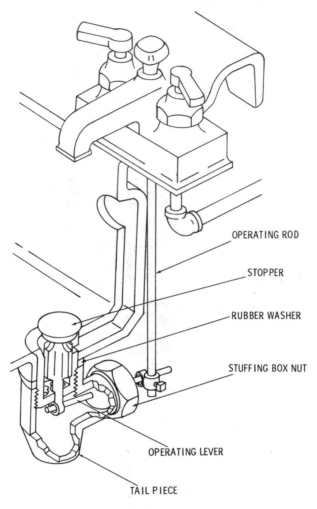

OPERATING ROD

STOPPER

RUBBER WASHER

STUFFING BOX NUT

OPERATING LEVER

TAIL PIECE

**Fig. 43. Details of a pop-up type drain.**

gives rapid automatic flushing and also provides an adjustment for the amount of water delivery. This type of valve (Fig. 44) is not normally used in residences, as the water delivery is rapid and noisy. It is recommended that a vacuum breaker be installed in the water supply line to prevent possible back siphonage in the event that reduced pressure in the water line occurs at the same

73

time that a stoppage or clogging in the water closet or urinal occurs. This could place polluted water in the supply line. The vacuum breaker prevents this possibility.

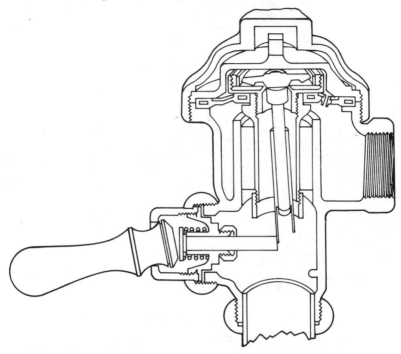

**Fig. 44. Cross section of a typical diaphragm-type flush valve.**

There is a variety of diaphragm-type flush valves manufactured, and it is therefore best to follow the manufacturer's recommendations when making repairs to this type of valve.

### Showers

Prefabricated steel-wall showers are the easiest to install, as most have prefabricated shower pans or a cement base is constructed into which the drain and trap are set. Custom-made tiled showers provide a longer life, but either type must have maximum-type waterproof walls and floors to insure a leak-proof installation.

Custom- or ceramic-type showers normally are installed with a lead shower pan into which the cement, tile, trap, and drain are

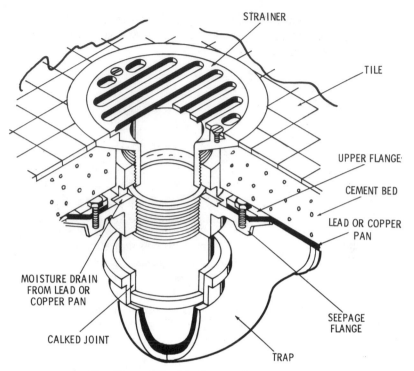

STRAINER

TILE

UPPER FLANGE

CEMENT BED

LEAD OR COPPER PAN

MOISTURE DRAIN FROM LEAD OR COPPER PAN

SEEPAGE FLANGE

CALKED JOINT

TRAP

**Fig. 45. Details of a shower drain installation.**

placed. A detailed cutaway view of a typical shower-pan and trap assembly is shown in Fig. 45.

Shower drains normally become clogged from an accumulation of scum and hair in the trap and occasionally in the waste line. Clear the stoppage by removing the drain strainer and using a force cup. If this fails, it may be necessary to clear the stoppage by the use of a trap-and-drain auger. If shower valves leak, they normally can be repaired by replacing the washers.

## Water Closets (Closet Bowls)

Closet bowls are manufactured in four general types—the *siphon-jet*, the *blow-out*, the *reverse-trap*, and the *washdown*.

**Siphon-Jet**—The flushing action of the *siphon-jet* bowl (Fig. 46) is accomplished by directing a jet of water through the up

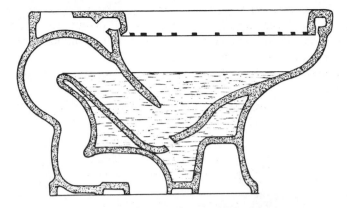

*Fig. 46. A siphon-jet type closet bowl.*

leg of the trapway, so as to fill the trapway and start the siphoning action instantaneously. The strong, quick, and relatively quiet action of the siphon-jet bowl, together with its deep water seal and large water surface, is universally recognized by sanitation authorities as the best type of closet bowl. These outstanding features make the siphon-jet bowl suitable for the most exacting installations.

**Blow-Out**—The *blow-out* bowl (Fig. 47) cannot be fairly compared with any other type because it depends entirely upon a

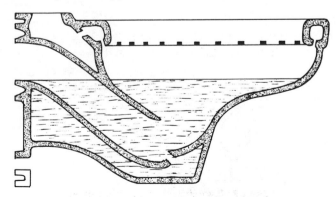

*Fig. 47. A blow-out type closet bowl.*

driving jet action for its efficiency rather than upon siphoning action in the trapway. It is economical in the use of water, yet has a large water surface that reduces fouling space, a deep water seal, and a large unrestricted trapway. Blow-out bowls are especially suited for use in schools, offices, and public buildings. They are operated with flush valves only.

**Reverse-Trap**—The flushing action and general appearance of the *reverse-trap* bowl (Fig. 48) is similar to the siphon-jet type. However, the water surface and size of the trapway are smaller and the depth seal is less, requiring less water for operation.

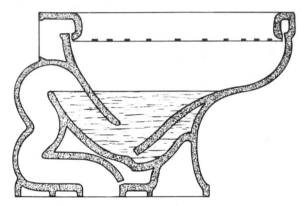

*Fig. 48. A reverse-trap type closet bowl.*

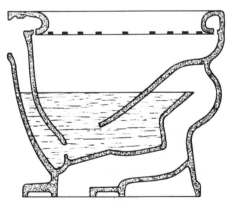

*Fig. 49. A washdown-type closet bowl.*

Reverse-trap bowls are generally suitable for installation with either a flush valve or a low tank.

**Washdown**—The *washdown* type bowl (Fig. 49) is simple in construction and yet highly efficient within its limitations. Proper functioning of the bowl is dependent upon siphoning action in the trapway, accelerated by the force of water from the jet directed over the dam. Washdown bowls are widely used where low cost is a prime factor. They will operate efficiently with either a flush valve or a low tank.

Stoppages in water closets are usually caused by foreign objects falling into the bowl. These obstructions can normally be cleared by using a force cup or, if necessary, a closet auger. If the stoppage is severe, and neither the force cup nor closet auger can remove the obstruction, it may be necessary to remove the closet bowl. To remove the closet bowl, refer to Figs. 50 and 51 and proceed as follows:

1. Shut the water off and empty the flush tank by sponging or siphoning.

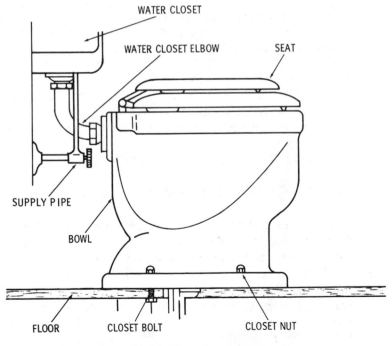

**Fig. 50. Closet-bowl details.**

GASKET

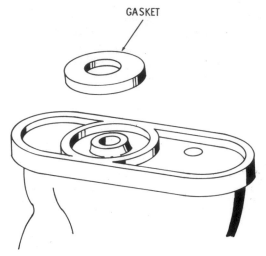

*Fig. 51. A bottom view of a typical closet-bowl base.*

2. Disconnect the water pipe from the tank.
3. Remove the tank from the bowl, if it is a two-piece unit, or disconnect the tank-bowl pipe connection if it is a wall-hung unit.
4. Remove the seat and cover from the bowl.
5. Remove the bolt covers from the base of the bowl and remove the bolts holding the bowl to the floor.
6. Break the seal at the bottom by jarring the bowl, and lift the bowl free.
7. Remove the obstruction from the discharge section.

To re-install the closet bowl, proceed as follows:

1. Obtain a wax seal or gasket from a plumbing supply house.
2. Clean the bottom of the bowl and place the wax seal or gasket around the bowl horn and press it into place.
3. Set the bowl over the soil pipe and press it into place.
4. Install the floor-flange bolts, drawing them up snug. Do not overtighten. To do so may crack the base of the bowl. Use a level (carpenter's type preferred) when tightening the bolts to make sure the bowl is level, using shims if necessary.

5. Re-install the items removed, including bolt covers, tank-pipe and water-pipe connections, seat, and cover.

Water-closet tanks (Fig. 52) are normally used in residential buildings or where quiet flushing is desired or necessary. A minimum amount of water is used during the flushing process. In locations where noise is not a factor, diaphragm-type flush valves are used. Although water closets may vary in design, their mechanisms are very similar.

**Float Valves**—If the float valve leaks, a worn plunger washer in the supply valve may be the cause. Shut off the water and drain the tank. Remove the screws that hold the levers, remove the plunger, and replace the water. If the entire float-valve assembly is badly corroded or seriously damaged by bending, replace the assembly.

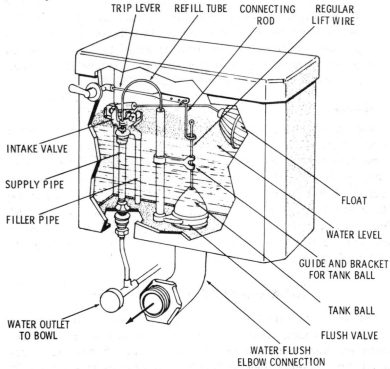

*Fig. 52. The flush-tank mechanism of a water closet.*

**Flush Valves**—Occasionally, the rubber ball of the flush valve becomes soft and loses its shape, resulting in poor base seating. If this occurs, replace the ball and, at the same time, remove all corrosion from the lift wire so as to prevent the wire from binding in the travel guide.

**Float Ball**—The purpose of the float ball is to maintain the water level in the flush tank at about 1 inch below the top of the overflow tube so as to provide sufficient water for a proper flushing action. If the ball develops a leak, it will not rise properly and the intake valve will remain open, causing a continuous flow of water. If the float ball cannot be drained and the leak soldered, replace the ball.

**Water-Closet Condensation**—Moisture condensing on the outside of the closet tank can be avoided by installing an insulating jacket which can be purchased at most plumbing stores. Also in use is a temperature device that mixes some hot water with the cold water entering the water closet. Installation of either device prevents condensation on the exterior of the water closet by keeping the temperature of the outer surface above the condensation or dew point of the surrounding air.

## Bathtubs

Almost all bathtubs presently being installed are of the built-in recessed type. There are square, rectangular, and angled tubs, some with ledge seats. Most are made of enameled steel or cast iron. In some residences, vitreous-china bathtubs have been installed. Most tubs are placed directly on the floor and, excepting those installed for a special purpose, require no wall support. Maintenance and repair generally consist of replacing valve washers and clearing stoppages. Most stoppages are cleared by the use of a force cup or a sink auger.

## THAWING FROZEN WATER PIPES

One of the best methods to thaw frozen piping is by the use of an electric heating cable. Even electric blankets have been used when no other means were available. Although thawing frozen water pipes is often done with a blow torch, this practice can be dangerous unless extreme care is taken to guard against starting a fire. A safer method is to cover the pipe with rags and then pour hot water over the covering.

It is recommended that a faucet be left open when thawing frozen water pipe. Start the thawing at or near the open faucet. Any steam resulting from the thawing process will escape through the open faucet and prevent the possible dangerous buildup of steam pressure in the pipe which could result in a pipe rupture or even an explosion.

## REPAIRING WATER VALVES AND FAUCETS

Many water faucets and valves are installed in the water system of an average building, and a knowledge of their repair is necessary. They must be inspected periodically for leaks, as leaking valves and faucets result in a loss to the building occupant or owner.

Most water faucets or globe-type angle valves need not be taken completely apart for repair. If the packing nut is removed, the stem or spindle can be taken out along with the washer assembly. Take the worn washer from the spindle and remove all of the bad or worn washer from the cup. Replace with a new washer. Do not use a washer that is too small—if necessary, file down a larger washer. If the seat of the faucet body is nicked or damaged, it can be refaced with a seat dressing tool that can be obtained or borrowed from most hardware or plumbing supply stores.

If the packing is in need of replacement, remove the handle, packing nut, and old packing, and install a new packing-type washer. If this type of washer is not available, stranded graphite-asbestos wicking works well by wrapping it around the spindle and turning the packing nut into place. In an emergency, even heavy cord coated with wax will often work.

Water faucets and globe valves (Fig. 53) are similar in design and leaks normally take one of two forms—a leaking washer or a leaking bonnet. Both forms of leaks can be repaired.

### Faucet Repair

To repair a faucet, shut off the water at the shutoff valve closest to the faucet being repaired. Take apart the faucet by removing the

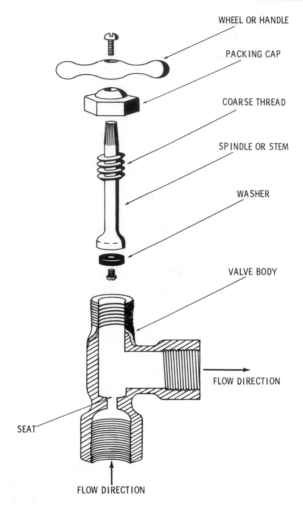

WHEEL OR HANDLE

PACKING CAP

COARSE THREAD

SPINDLE OR STEM

WASHER

VALVE BODY

FLOW DIRECTION

SEAT

FLOW DIRECTION

*Fig. 53. A globe-type angle valve.*

handle, packing nut, packing, and stem in that order. The handle may be used to unscrew and remove the stem.

After removing the screw and worn washer from the stem install a new washer of the proper size and type and reassemble the faucet.

Mixing faucets, which are found on sinks, laundry trays, and bathtubs are normally two separate units with a common spout. Repair each unit independently. A noisy faucet when the water is running may be the result of a loose washer or worn threads on the stem, permitting the stem to chatter or vibrate. Pressing down on a bad stem will stop vibration but will not affect a loose washer.

In most cases new stems are available from supply houses, as are stem receivers and seats. If the stem, stem receiver and the seat of the stem have been replaced then all wearing parts of the faucet have been replaced.

Often, when a shower head drips the supply valve has not been fully closed. If completely shutting off the valve does not stop the drip, the valve must be repaired.

## Repairing Pipe Leaks

Leaking of a threaded pipe connection can generally be stopped by unscrewing the fitting and applying a pipe-joint compound which seals the joint when it is screwed back together.

Small pipe leaks can be repaired with a rubber patch and sleeve or metal clamp. This type of repair is only an emergency repair and permanent repairs must be made as soon as possible.

Large leaks require the cutting out of the bad section and the installation of a new piece of pipe. Unless the leak is near the end of the pipe, the installation of a pipe union will be required. Temporary repairs can be made with plastic or rubber tubing which must be strong enough to withstand the water pressure in the pipe. To make the repair slip the open ends of the plastic or rubber piping over the leak section and fasten with several turns of wire or with pipe clamps.

In copper water piping, vibration may result in a leak. When such leaks occur DRY, CLEAN, and RESOLDER the joint.

## TYPICAL FROSTPROOF HYDRANT

When water is left in frostproof hydrants it soon runs out of the drain tube (Fig. 54) which prevents water from freezing in the hydrant during cold weather. There are two important features in frostproof hydrants (which are basically faucets) and these are:

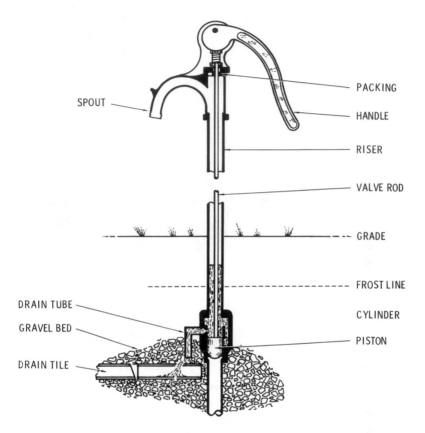

SPOUT

PACKING

HANDLE

RISER

VALVE ROD

GRADE

FROST LINE

DRAIN TUBE

CYLINDER

GRAVEL BED

PISTON

DRAIN TILE

*Fig. 54. Typical frostproof hydrant.*

(1) the valve is designed to drain the water from the hydrant when the valve is shutoff and (2) the valve is installed underground, below the frostline to prevent freezing. As is true of most ordinary faucets worn packing, gaskets and washers often result in leakage. To repair, disassemble the hydrant and replace or repair worn parts. It is recommended that these hydrants should be used only where positive drainage can be provided since the water flowing from the hydrant may draw in contaminated water standing above the hydrant drain level causing water contamination.

## GREASE TRAPS

Where large quantities of grease are disposed of through sink drains, particularly in restaurant kitchens, it is necessary to install a grease trap in the waste line. The grease trap should be installed as close as possible to the main or prime sink, such as

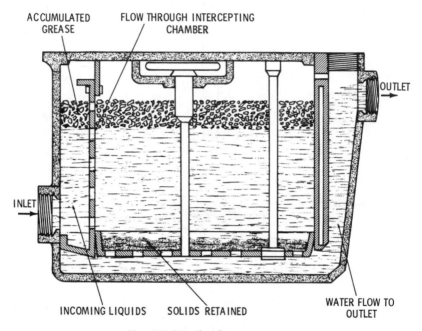

**Fig. 55. Details of a grease trap.**

the scullery sink. If grease is allowed to pass into the waste pipe, clogging and stoppage of water flow through the system will eventually occur.

Grease traps vary in their design, but all operate on the principle that grease is lighter than water and will therefore rise to the top as it goes through the grease trap and cools. The water passes through the trap, but the grease rises to the top and accumulates. A cross-section of a typical grease trap is illustrated in Fig. 55.

During excessive kitchen use, the trap may fill rapidly with grease and will not operate to remove additional grease until the accumulated grease is removed. Grease traps are nearly maintenance free, but the accumulated grease must be periodically removed.

## SUMP PUMPS

Sump pumps are installed in locations which are below the sewer-level waste line. An electrical centrifugal-type sump pump is normally used for this type installation and is equipped with switches, floats, and controls which permit its automatic operation.

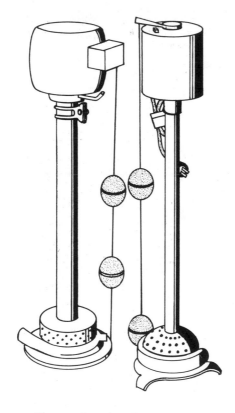

***Fig. 56. A pedestal-type sump pump.***

Maintenance consists primarily of periodic checks to clean the screens and to assure that the motor and bearings are lubricated, either by means of attached grease cups (the grease cup cap must be slightly turned) or by oiling; the procedure depends on the type or manufacturer of the unit. Repairs to the unit must be made in accordance with the instruction booklet furnished by the manufacturer.

Two types of sump pumps are in normal use—the *pedestal type* (Fig. 56) where the electric motor is above the floodwater line, and the *submersible type* (Fig. 57) where the entire unit is below the possible high-water line.

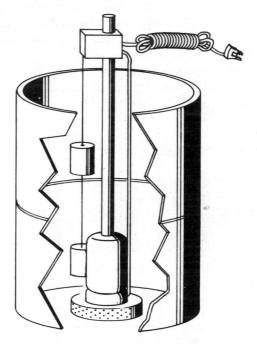

**Fig. 57. A submersible-type sump pump.**

## AIR AND GAS PIPING

### Air Piping

Air piping is usually installed for special uses, such as in garages, service stations, machine shops, general service shops, and laundries, or in any location where compressed-air systems are a requirement.

Compressed air is obtained either by a diesel, gasoline, or electrically driven compressor, and is stored in a compressor tank for immediate or future use. As the unit operates, air is drawn into the compressor, is reduced in volume, and goes through a check valve into a tank for storage.

The compressor operates automatically through a combination pressure-control and safety valve, and is shut off when the pressure in the storage tank is at a desired level. If the pressure-control valve should fail, the safety valve in the tank will relieve the tank pressure. A gauged reducing valve is used to draw the compressed air from the tank, and this valve maintains almost any desired pressure to the use source.

When installing air piping, it is recommended that all sharp bends be avoided to reduce possible friction and air-pressure loss. Recommended safety practices include adequate pressure-relief valves to be installed on all air compressors, and all pressure to be removed from the lines before working on them. A typical air-compressor installation is shown in Fig. 58.

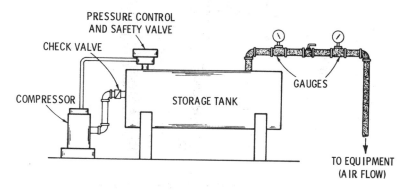

**Fig. 58.** *Details of a commercial air-compressor installation.*

## Gas Piping

Natural or illuminating gas is almost odorless. Before distribution to the consumer, odorous chemicals are added so that the consumer can detect any leaking gas by smell. Steel or black-iron pipe with threaded joints is generally used for gas services. Gas contains considerable moisture, so the piping should be sloped to accessible points where drip cocks are installed to drain off moisturized condensate in the lines. Gas piping, unions, and bells should be left exposed, where possible, for ease of inspection.

Gas piping on the exterior of buildings is normally laid underground and is normally welded at the joints. Threaded pipe reduces the thickness of the pipe walls at the joints by approximately 35 per cent, which permits deterioration more quickly by corrosion. It is recommended that all gas piping be installed by the gas company employees, when possible, or by men who have a good working knowledge of gas piping.

Some safety precautions to follow when working on gas lines are as follows:

1. Smoking should not be permitted in areas where gas piping is being installed or repaired.
2. Gas piping should not be tested for leaks with any flame.
3. Pressure should be removed from gas lines before working on them.

# CHAPTER 3

# Concrete and Masonry

The term *concrete* can be defined as a mixture of water, cement, and aggregate mixed together to form a paste. When this paste dries, an artificial stone is formed. If concrete is poured into a mold while still a paste, it will assume the shape of the mold or form when it hardens.

A basic requirement of good concrete work is proper mixing and proportioning of the ingredients. Cement is generally referred to as *Portland cement*, which is an artificial mixture of lime and clay-bearing materials which are burned in a kiln to a point of fusion and then ground to a fine powder. It is said that Portland cement got its name because, when it was patented in the early 1800's, the finished product had the hardened appearance of rock quarried on Portland Island in the English Channel. Aggregate is classified as fine or coarse. Coarse aggregate refers to either gravel or crushed stone. Fine aggregate is sand.

In many instances, concrete work requires the use of reinforcing steel rods or mesh. When rods or mesh are imbedded in the concrete, it is referred to as *reinforced concrete*. Concrete normally resists compressive stresses. When bar reinforcing steel is imbedded in concrete, it is assumed it will provide for tension stresses.

## CEMENT STORAGE

Cement should be stored on platforms built approximately 12 to 18 inches above the ground for purposes of ample ventilation and circulation of air. Cement must be kept dry at all times. During wet or damp periods, cement may become hardened from air moisture, causing it to set. If this occurs, the cement should not be used.

## CONCRETE MIXES

A common term used in concrete work is the *mix*. Since the type of work for which concrete is to be used is not always the same, the proportioning of the ingredients—cement, water, and aggregate—will vary. The *mix* determines the consistency and compressive strength of the finished concrete. Therefore, proper proportioning of the ingredients is required.

Concrete is generally available for use in three ways—buying the ingredients separately and adding water, premixed concrete delivered by concrete-mixing trucks for use when laying foundations, driveways, patios, and other large concrete jobs, and by buying premixed materials by the bag which requires only the addition of water.

No matter how the concrete ingredients are obtained, *do not add too much water*. This prevents the various components from staying together properly and weakens the finished product. Concrete with too much water in it hardens into a solid state, but it may be only 60 per cent as strong and durable as concrete with the right amount of water in it.

If the instructions on how to make concrete indicate a 1-2-3 mix, it means 1 part Portland cement, 2 parts sand, and 3 parts aggregate (usually gravel). The same thing holds true regardless of which numbers are used. For instance, a 1-2½-3½ mixture means 1 part Portland cement, 2½ parts sand, and 3½ parts gravel.

The water-cement ratio for normal building construction in moderate climates should not exceed 6 gallons of water per bag of cement. After the water-cement ratio has been selected, the proportions of aggregates and cement plus water should be adjusted to produce a mixture of proper consistency. For concrete in severe climates, the water-cement ratio should not exceed 5 gallons of water per bag of cement. The proportions listed in Tables 1 and 2 for concrete mixes are suggested as suitable under normal or average conditions.

CONCRETE AND MASONRY

## Hand-Mix Concrete

It is not recommended to hand mix concrete if the batch is over 1/2 cubic yard, or is of such a volume that it cannot be put into place in a 1/2-hour period or less. Machine mixing for almost all work, excepting small maintenance jobs, is recommended. When hand mixing, use a tight wooden platform, concrete floor, or a similar surface. A 10 square foot surface is about the right size for hand mixing single batches of concrete.

The following procedure for hand mixing is generally used for small batches:

1. Spread a measured amount of sand and aggregate for the batch 3 or 4 inches deep on the platform.
2. Spread the premeasured cement over the sand and mix the two together with a hoe or shovel until the mix has an even gray color.
3. Add about half the measured amount of the required water and mix thoroughly until a smooth and an even gray mortar is obtained.
4. Spread this mortar on the batter board. Slightly wet the coarse aggregate and add to the mortar, turning the mass

## Table 1. Concrete Mix Proportions for Average Conditions

| Type of Work | Slump | Mixture | Maximum Size Aggregate | Water Ratio (gallons per sack) |
|---|---|---|---|---|
| Barn approaches | 2"-3" | 1-2-3½ | 1½" | 6 |
| Areaways | 2"-3" | 1-2-3½ | 1½" | 6 |
| Walls | 2"-3" | 1-2-3½ | 1½" | 5.5 |
| Walls subject to moisture | 2"-3" | 1-2-3 | 1½" | 4.75 |
| Septic Tanks | 3"-5" | 1-2-3½ | 1" | 5.5 |
| Steps and Stairways | 2"-4" | 1-2-3½ | 1" | 6 |
| Sidewalks | 2"-4" | 1-2½-4 | 1½" | 6 |
| Retaining Walls | 3"-5" | 1-2-3½ | 1½" | 5.25 |
| Floors, Plain | 2"-4" | 1-2½-4 | 1½" | 6.25 |
| Floors, Reinf. | 3"-5" | 1-2-3 | 1" | 4.50 |
| Fence Posts | 2"-4" | 1-2-3 | ¾" | 4.50 |
| Driveways | 2"-4" | 1-2-3½ | 1½" | 5.25 |
| Curbs | 4"-6" | 1-2-4 | 1½" | 6.25 |
| Basements | 3"-5" | 1-2-4 | 1½" | 5.50 |
| Boiler Settings | 4"-6" | 1-2-3½ | 2" | 5.25 |

## Table 2. Proportioned Concrete Mix by Weight

| Cement-Water Ratio | Maximum Gravel Size | POUNDS OF DRY AGGREGATE | | | |
|---|---|---|---|---|---|
| | | Round | | Angular | |
| | | SAND | GRAVEL | SAND | GRAVEL |
| 1 bag cement 5 gal. water | ¾" | 182 | 230 | 182 | 182 |
| | 1  " | 172 | 250 | 170 | 205 |
| | 1½" | 170 | 300 | 170 | 240 |
| | 2  " | 170 | 335 | 170 | 275 |
| 1 bag cement 6 gal. water | ¾" | 235 | 275 | 235 | 225 |
| | 1  " | 225 | 305 | 225 | 250 |
| | 1½" | 220 | 350 | 220 | 290 |
| | 2  " | 220 | 410 | 220 | 335 |

with a shovel or hoe, adding more water (premeasured) until a good plastic mass is obtained. The mass should be plastic, but not so sloppy that it runs off the shovel.

Some concrete workers prefer to hand mix their cement by dry mixing the cement and sand. If this procedure is used, spread the cement and sand (premeasured) in a circle and then add the coarse aggregate and water, thoroughly mixing into a good plastic mass. This method also works well for small batches of hand-mixed concrete.

### Ready-Mix Concrete

Almost all ready-mix concrete is of a good quality. Tests have indicated, however, that in some instances the strength was less than that for the stationary mixing-machine type. In these instances, it was revealed that the poorer quality mix was due to unreliable mixing personnel and obsolete truck-mixing equipment.

Ready-mix concrete batches should have not less than 50 nor more than 100 revolutions at a mixing speed not less than 4 rpm after all materials are in the mixer drum. The capacity of the individual mixer and the rotating speed should be as recommended by the manufacturer.

### Conveying and Placing

Concrete should be conveyed as quickly as possible from the mixer to the forms to prevent segregation of materials. Place the concrete paste into all angles and corners of the forms and around the reinforcement. Under normal conditions, not more than 30 minutes should elapse after mixing to place the concrete into the forms. Placing of the concrete should be carried on so that pouring over cold joints is prevented. If circumstances do not permit this, then the cold joints should be scarified and broomed with a thin coating of cement before beginning the joining pour. Do not place concrete on ice, standing water, dirt, heavy dust, or other foreign matter.

Concrete should be deposited in the forms in the final position to avoid rehandling, if possible. Concrete that has partially hardened should not be deposited in the work. When concrete work is started, the pouring and placing should be continuous until an entire section is completed, with the work being stopped at locations which will not weaken the structure.

## SIDEWALKS

Concrete for sidewalks may be placed directly on a well-tamped or firm soil foundation if there is good natural drainage. However, a 4-inch base of packed gravel is recommended. Pitch of the finished surface, from one side to the other, should be a minimum of 1/4" to 1/2". The finished sidewalk surface should be at least 1 inch above the surrounding grade so as to prevent an accumulation of water on the sidewalk.

## GARAGE FLOORS

Two-car private-home garage floors, and small building floors, are usually laid directly on packed and firm earth excavated to the required depth, but without a base or foundation. However, 4 inches of packed gravel or coarse sand is recommended for the base. Because concrete will pick up moisture, the general practice is to lay a thickness of sheet plastic or tar paper on the base. The concrete is then poured directly on the moisture-barrier material.

## UNDERWATER CONCRETE

Concrete can be placed under water by filling burlap sacks with concrete and then lowering the sacks into the water. Funnel-like forms can also be used, as indicated in Fig. 1.

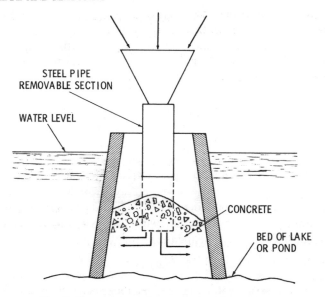

STEEL PIPE
REMOVABLE SECTION

WATER LEVEL

CONCRETE

BED OF LAKE
OR POND

*Fig. 1. Concrete can be poured under water.*

Low-slump (stiff) concrete is recommended for underwater placement. A 1-1/2" to 2-1/2" diameter blunt-end pole can be used for compaction purposes. Every precaution must be taken in this type of placement to prevent the cement from being washed away from the aggregate. A fairly rich mix, containing at least seven sacks of cement per cubic yard of concrete, should be used.

## CONCRETE CURING

### Curing With Fabric

Keep fabrics such as burlap, cotton-quilt type, etc., continuously wet during the curing period.

### Curing with Saturated Sand

Cover the concrete surface with about 1 inch of evenly spread sand and keep saturated continuously during the curing process.

### Curing with Paper

An asphalt-impregnated or an aluminum-foil paper is often used for curing purposes. If paper of this type is used, lap the ends about 4 inches and tape or use waterproof cement to seal the joints.

## Protection from Freezing

Protection from freezing can be accomplished by covering the concrete on ground surfaces with straw, dirt, canvas sheets, or other such type materials. Protection should be continued for at least five days.

All dirt subgrades over which concrete is to be placed must be protected with canvas sheets, straw, hay, or other such materials. The covering should be placed after completing the excavation and should not be removed until the concrete is ready to be poured. Concrete must not be placed over frozen subgrade, as this may result in wall cracks by settlement after the subgrade thaws.

## Excavation Depths

If no excavation depths have been determined for footings, the depths listed in Table 3 may be used as a guide for average soil conditions.

### Table 3. Excavation Depths for Concrete Footings

| Mean Temperature Zone (°F.) | Minimum Depth in Feet Below Existing Grade |
|---|---|
| Plus 20 degrees | 2 |
| 0 degrees | 3 |
| Minus 20 degrees | 4 |

Care must be used not to excavate below the suggested depths, but if it does happen, pour the concrete into the excavated depths. Do not fill the extra excavated depth with soil and try to compact it, as getting good compaction is difficult, and poor compaction could result in settling that may crack the footings and building walls.

## Shoring

If soil and wet-weather conditions warrant, shoring of the deeper trenches may be necessary. Fig. 2 shows typical shoring and bracing methods being used.

In order to protect personnel and property, it is necessary that excavations and open trenches have railings and proper night-time lighting when working personnel are not actually on the job site. Safety provisions for excavations, trenching, and trench-and-bank shoring are a necessity.

## CONCRETE FORMS

Concrete forms must be strong, rigid, economical, almost water-tight, and simple. Therefore, the materials most commonly used for this purpose are wood and metal. Metal forms can be used over and over again, but for small construction jobs and repair work, wood forms are generally used because of economy.

### Wood Forms

An exterior-grade plywood or common lumber can be used for concrete forms if it is of the required thickness and strength to perform the intended job. Before using the wood, large checks, splits, loose knots, or other defects that may affect the appearance of the finished concrete should be repaired. Metal pieces can be fitted over the defects if they are not too large. If the defects are too large, discard the particular piece of lumber.

Tongue-and-groove lumber can be used, but in most forming, exterior-grade plywood is used. Lumber used for forming is generally oiled so that the concrete surface is not spalled or roughened when the forms are removed. Forms, however, should not be oiled if the concrete surface is to be painted or plastered. Almost all of the lighter petroleum oils are satisfactory for concrete form use. Waste oil or those containing too much sludge often stain the concrete and are not satisfactory excepting for use on exterior concrete surfaces that will be covered by soil. If wood forms are not treated with oil, they should be completely wetted with water before the concrete is poured into place.

### Metal Forms

In oiling metal concrete forms, a compound oil is generally used. Although these oils have a petroleum base, they also contain animal or vegetable oils and are better for metal forms than common petroleum oil.

In removing concrete forms under normal weather conditions, the average time schedule listed in Table 4 may be used. This schedule cannot be applied in all cases, but it is an average. Remember that concrete only a few days old is still relatively weak and may crumble along the edges if the forms are removed too soon. When possible, leave the forms in place a few extra days.

## Table 4. Concrete Form Removal Schedule

| Type of Work | Above 60°F. | 50°F. to 60°F. | 35°F. to 50°F. |
|---|---|---|---|
| Footing forms | 2 days | 5 days | 8 days |
| Wall and column forms | 5 days | 7 days | 9 days |

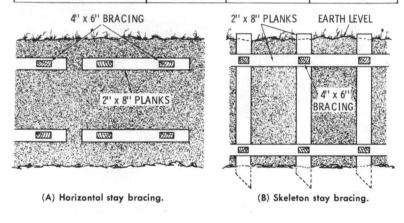

(A) Horizontal stay bracing.  (B) Skeleton stay bracing.

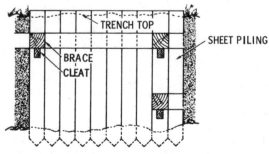

(C) Sheet piling method.

**Fig. 2. Types of shoring used on excavations.**

For sidewalks, 4-inch floors, and other similar concrete construction, $2'' \times 4''$ forms with stakes driven along the outside

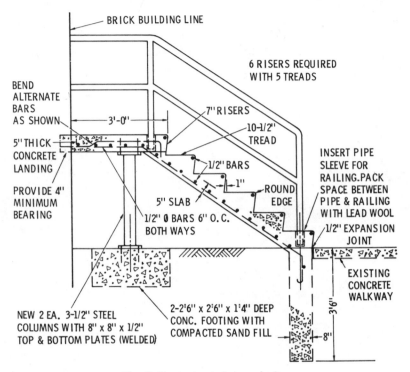

*Fig. 3. Concrete stairway design.*

for anchoring is sufficient. The $2'' \times 4''$ forms should be placed in such a manner that the top edges will act as guides for the finished level of work. Leave the forms in place until the concrete will not crumble when they are removed.

Typical forms for steps, walls, curbs, and gutters are shown in Figs. 3, 4, 5, 6, and 7. Common cement finishing tools are shown in Figs. 8, 9, 10, and 11.

## CONCRETE PATCHING

When making minor repairs on sidewalks, patios, and concrete floors, a ready-mix concrete packaged product may be used.

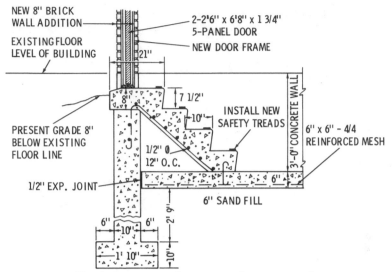

NEW 8" BRICK WALL ADDITION

EXISTING FLOOR LEVEL OF BUILDING

2-2'6" x 6'8" x 1 3/4" 5-PANEL DOOR

NEW DOOR FRAME

21"

8"

7 1/2"

10"

INSTALL NEW SAFETY TREADS

3'-0" CONCRETE WALL

6" x 6" - 4/4 REINFORCED MESH

PRESENT GRADE 8" BELOW EXISTING FLOOR LINE

1/2" ∅ 12" O.C.

1/2" EXP. JOINT

6"

6" SAND FILL

6"

6"

2' 9"

10"

10"

1' 10"

**Fig. 4. Concrete stair to a step-down areaway.**

This mixture requires only the addition of a proportioned amount of water. Cracks must be cleared of all loose fragments and plant life, and then wetted thoroughly with water. The concrete paste is then packed into the openings and smoothed down. The patches should be moistened with water for about 4 days to prevent too rapid drying.

Large cracks may be patched by cutting out the cracks to a depth of at least 1 inch and a minimum width of 1/2 inch. Wet down the entire opening with water and fill with a 2-to-1 sand-cement mortar mix.

For larger cracks, including those in reinforced concrete walls, cut out the damaged areas so as to expose any reinforcement, and then clean thoroughly. Wet the surface completely with water and allow to dry about 1 hour. About 20 minutes before the patching cement is to be applied, paint the entire area to be patched with water and cement mixed to a consistency of thick cream. A broom may be used for brushing the cement slurry into the patch area. Do not use this slush coating if the cement removed is down to the fill area of the work.

For all patching work, keep the water in the mix at a minimum, using not more than 4 gallons for each sack of cement. Tamp

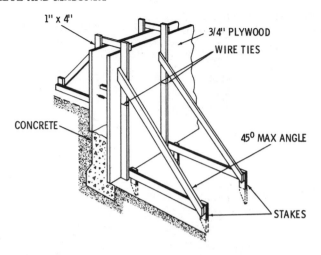

1" x 4"

3/4" PLYWOOD

WIRE TIES

CONCRETE

45° MAX ANGLE

STAKES

*Fig. 5. Concrete forms for walls in firm ground.*

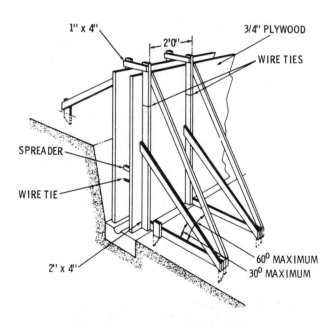

1" x 4"

2'0"

3/4" PLYWOOD

WIRE TIES

SPREADER

WIRE TIE

2" x 4"

60° MAXIMUM

30° MAXIMUM

*Fig. 6. Concrete forms for wall in soft ground.*

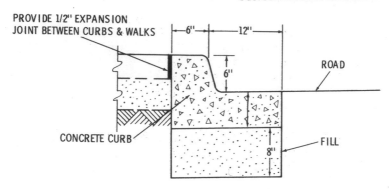

PROVIDE 1/2" EXPANSION
JOINT BETWEEN CURBS & WALKS

6"  12"

6"

ROAD

CONCRETE CURB

FILL

8"

Fig. 7. Cross section of a typical concrete curb and gutter.

Fig. 8. A standard cement
finishing trowel.

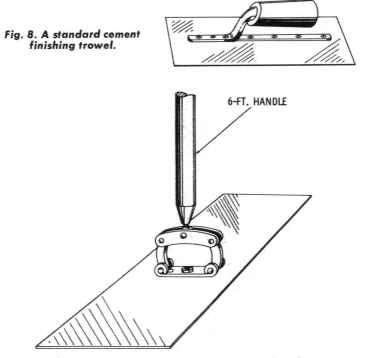

6-FT. HANDLE

Fig. 9. A "push-pull" cement finishing trowel with a
long handle. A double-action device tilts the
leading edge up when the trowel is
either pulled or pushed.

103

the patch material into place, screed and float the surface to the required level, let stand for about 35 minutes or until pressure from the fingers or hand does not dent the fresh concrete, and then steel trowel to a complete finish. Existing expansion joints should be maintained. Screeding (leveling) is generally accomplished with a straight-edge board.

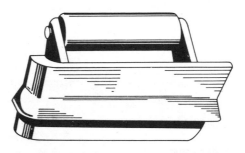

*Fig. 10. A cement groover used in finishing cement walks and driveways.*

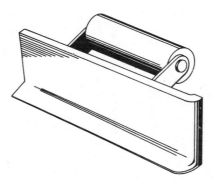

*Fig. 11. A cement edger used to "round-off" the corners and edges of steps, sidewalks, and driveways.*

Shrinkage cracks are fine hairline cracks that appear in concrete. In repairing these cracks, do not chisel out and fill them. Instead, scrub such cracks with a grout made with 65 per cent cement and 35 per cent fine sand, mixed to a syrup-like consistency. Wet the area and scrub in the mixture. It is not recommended that shrinkage cracks be repaired until the concrete is at least 1 year old. However, if the concrete is exposed to severe weather, such as heavy rains, corrosion, or sea water, these cracks should be repaired as soon as possible.

## BRICK AND CONCRETE MASONRY

### Brick

Many types of brick are manufactured, but the one most used is the *common brick*. It is generally made from surface shale or clay, with the materials mixed, shaped, and baked in a kiln. Generally, common brick are those cut from a mold or column of a clay mixture with a wire or blade at the factory. *Smooth* or *face brick* are those which are made in trays or in individual molds.

Another type of brick is the high-temperature *firebrick* used for the interior of smoke stacks, flues, fireplaces, furnaces, and chimneys. *Glazed* and *enameled brick* are used for sanitary purposes in locations such as kitchens and hospitals. Common brick are mostly used for building structures and walls, *paving brick* (which are normally hard) for paving roads and streets, and *pressed brick* are used where color or appearance requires a brick that is more suitable for the selected work.

The standard sizes for brick commonly used are listed in Table 6.

In addition to the commonly used brick sizes, larger size brick are being manufactured in some sections of the country. Producers say the larger size building brick reduce labor installation time and thus lower building costs. In addition, they are more suitable for certain types of architectural treatment.

### Table 6. Sizes of Standard Brick

| Type of Brick | Dimensions (in inches) | | |
|---|---|---|---|
| | Width | Length | Depth |
| Rough Face | 3¾ | 8 | 2¼ |
| Smooth Face | 3⅞ | 8 | 2¼ |
| Common | 3¾ | 8 | 2¼ |

### Structural Clay Tile

Clay tile used for building purposes is generally hollow and is available in many different finishes, sizes, and shapes. It is available in the common brick sizes and up to 8 × 16 inches, and in thicknesses varying from 2 to 12 inches. They can be obtained in many shapes and several finishes. Included in the finishes are *salt glaze* and *ceramic glaze*.

### Concrete Masonry

Concrete masonry blocks, in general use, are made with either light-weight or with heavy-weight aggregate. The light-weight, hollow, load-bearing types are made with coal cinders, expanded shale, clay, or slag, and other materials such as volcanic cinders and pumice. Heavy-weight units are composed of sand, gravel, air-cooled slag, and/or crushed stone. In both types, Portland cement is proportioned in accordance with tested requirements. A standard 8 × 8 × 16-inch light-weight unit will weigh from 25 to 35 pounds, depending on the base ingredients. The heavy-weight type will weigh from 40 to 50 pounds each. These units are used in all types of construction, depending on the job and design requirements.

In addition to concrete blocks, concrete building tile and concrete brick are available for construction and general repair purposes. Concrete masonry (blocks and units) should *not* be wetted when placing, but should be laid dry.

Concrete block and special concrete units are available in various sizes, but the size most commonly used is the 8 × 8 × 16-inch unit which occupies the same wall space as would 12 standard bricks and their mortar joints. Because of their size, concrete block units are installed more rapidly than the smaller brick masonry type. Further simplifying the installation of concrete blocks are the many special units which include half and quarter sizes, and corner and jamb units.

### Mortar

A good mortar mix is essential in producing durable weather-resisting construction. It is recommended that salt or chemicals that retard freezing not be added in the preparation of mortars.

Mortar strength may be reduced by the additive. Added salts may result in efflorescence (efflorescence is a white deposit of soluble salts which appear on the surface of walls, detracting from their appearances).

Straight cement mortar (1 part Portland cement and 3 parts sand) offers great resistance to compression and is generally used where dampness exists or may be encountered, such as in basement walls.

Cement-lime mortar (1 part Portland cement, 1 part lime putty, and 6 parts sand) sets more slowly than straight cement mortar and offers a more complete bond. This type of mortar is generally used for brick and stone work.

A partial list of mortar mixtures which are used in normal masonry construction is given in Table 7.

For tuckpointing purposes and small masonry repairs, it is recommended that special mortar mixes which are sold under various trade names be purchased. These mixes carry specific recom-

### Table 7. Mortar Mixes

| Mortar Type | Parts by Volume | Material |
|---|---|---|
| Masonry cement | 1<br>3 | Masonry cement<br>Sand |
| High Portland cement | 1<br>3 | Portland cement<br>Sand |
| Lime cement | 1<br>1<br>6 | Lime-putty<br>Portland cement<br>Sand |

mendations by the manufacturers as to their use, and the proper mixture for the intended use is assured.

## Weather Conditions

Masonry structures can be and are being built in freezing weather. All materials should be heated so they will remain above the freezing point (32°F) until after working into place. Finished masonry should be protected from freezing for at least 2 days.

In mild weather (over 40°F), common or pressed brick should be laid damp so as to prevent future hairline cracks between the brick and the mortar. Bricks should be wetted with water for at least 45 minutes before laying. However, no surface water should be on the brick when it is placed.

## Tuckpointing

Water leaks on the interior of building walls often are the result of deteriorated mortar at brick and stone joints. When making mortar-joint inspections, look for hairline cracks where the mortar has pulled away from the brick and also crumbling of mortar which results in gaps between the brick.

Generally used for preparing the joints for repair are a hammer and cold chisel. Care should be used in chipping out loose or deteriorated mortar so as not to damage the firm joints and bricks. Chip away all old mortar to a depth of at least 3/4 inch. If the repair work is minor, a ready-mixed mortar can be purchased. Using the mixed mortar and a pointing trowel, work small amounts of the material into the sections being repaired, after wetting the surface with water. Avoid smearing the face of the brick. Fill the openings flush with the brick, allow to dry for a short time (until firm to the touch), and then draw the trowel across the joint leaving a V groove. Keep wet at least overnight to keep the mortar from drying out too fast (wetted batting with a braced board is one method used to keep the joints moist).

Entire bricks can be replaced, using the same brick if it is not damaged. Clean the brick and opening thoroughly, wet both opening and brick with water, apply mortar to the brick and press it into the opening, filling the voids with mortar. This general method of pointing applies to all brick work.

## Joints

Good mortar joints, with proper workmanship, are necessary to produce a masonry wall that will serve its intended purpose—that of strength, resistance to water, good appearance, and durability. In normal construction, 3/8-inch joints are best for making brick bonds, and 1/4- to 3/8-inch joints for concrete masonry.

A *struck joint* (Fig. 12) is normally used on interior walls. It is not recommended for exterior walls, as the small shelf houses moisture and may cause damage to the mortar, particularly if hairline cracks appear.

A *weathered joint* (Fig. 13) is efficient and water shedding for exterior work, as no shelf is left on the lower brick by the mortar joint.

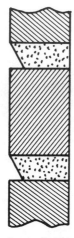

Fig. 12. A struck mortar joint.

Fig. 13. A weathered mortar joint.

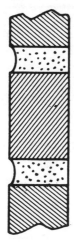

Fig. 14. A concave mortar joint.

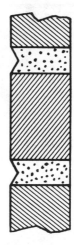

Fig. 15. A V mortar joint.

*Concave* and *V-joints*, sometimes called *tooled joints* (Figs. 14 and 15), are desirable as they are considered inexpensive to make and are resistant to inclement weather.

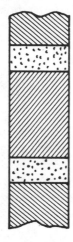

**Fig. 16. A flush or plain-cut mortar joint.**

A *flush* or *plain-cut joint* (Fig. 16) is made by cutting the mortar flush with the wall surface. This type of joint is not considered to resist water entrance as well as the tooled joints.

## GLASS BLOCKS

Glass blocks are generally of three types—*decorative,* where design of the block is the main consideration; *functional,* which controls the light rays as they pass through the glass; and *general-purpose,* which diffuses the light without direction control.

Glass blocks are made by fusing two hollow sections together under high temperature. They are not load bearing and should not be subjected to other than their own weight when used for construction. They are practically maintenance free and require only an occasional cleaning.

The sections of the blocks which receive mortar can be smooth or corrugated. These surfaces normally have a factory applied grit which forms a mortar bond.

### Mortar

Glass-block mortar is usually made of Portland cement Type 1, lime or quicklime, and fine sand. The sand should be of a type where not more than 12 per cent by weight passes a No. 100 sieve and all or 100 per cent passes a No. 8 sieve.

When mixing the mortar, use 1 part lime putty, 4 parts sand, and 1 part Portland cement. The mortar should have a stiff workable consistency. In using lime, the following procedure should be used. Soak hydrated lime at least 12 hours; slake quicklime and store for a minimum of 20 hours, or until cooled. Screening is necessary before its use.

## Installation

Glass blocks are normally laid in a checkerboard pattern with all blocks plumb and true. Recommended mortar-joint widths are 1/4″ and 3/8″ with a concave-tooled joint. Reinforcing ties should be continuous from one to the other end of the panel. Ties should be bedded entirely in the mortar, but should not be allowed to bridge the expansion joints. Expansion joints not less than 1/2-inch wide should be provided at the jambs and head of every glass-block panel, and intermediate expansion joints provided if the panels are over 20 feet in length or width. These joints should be packed with oakum or filler strips of the premolded expansion type. The panels should be calked to a minimum depth of 1/2 inch with a good calking compound around the interior and exterior perimeters.

## Repair and Maintenance

Glass blocks require a minimum amount of maintenance other than a periodic inspection of the joints and cleaning of the block surfaces at desired intervals. Glass blocks may occasionally crack or break, and when this happens, the following replacement procedure should be used:

Chip off all broken glass clinging to the undamaged portions, using care not to damage the wall ties, expansion joints, or the other glass blocks. Carefully clean all old mortar from the exposed panels and install a new glass block with mortar, making sure the work is true and level. Ready-mixed commercial mortar materials are available for repair jobs of this type. Recalk as necessary. If ready-mixed mortar material is not available, use 1 part Portland cement and 2 parts of fine sand mixed to a thick consistency. Mix only small batches.

## BATHTUB AND SHOWER CRACK REPAIR

If cracks appear between the bathtub and wall, repair these cracks as soon as possible to prevent water damage to the wall and building frame.

Cracks of this type can be repaired with waterproof grout or a plastic sealer.

Grout is sold in a powder form and must be mixed with water to form a thick paste. Mix in small amounts to provide only the amount required for the particular job.

Press this mixture into the crack with a putty knife, then smooth the surface. Wipe away any excess grout from the wall and tub before it drys. Do not use the tub until the grout in the repaired crack has dried.

Plastic sealer, which costs more than the grout, comes in a tube like toothpaste and is easier to use than the tub and shower area crack repair. Follow the tube listed manufacturer's directions when applying.

Squeeze the plastic sealer from the tube into the crack, smooth it with a putty knife cleaning away any excess sealer material. Work fast as the average plastic sealer dries quickly.

## CONCRETE AND MASONRY STAIN REMOVAL

**Tobacco Stains**—Minor tobacco stains can be removed by the application of a 1/4- to 1/2-inch thick paste made from water and any commercial grit scrubbing powder. Allow to dry, scrape off the paste, and wash the surface with clean water. For heavy or stubborn tobacco stains, mix 10 ounces of chlorinated lime with water in a dish or enameled pan to form a paste. Place 2 lbs. of trisodium phosphate crystals in 4 quarts of warm water and allow to dissolve. Pour both mixtures into a 2-gallon container, fill with water, and allow the lime to settle. Add some of this liquid to powdered talc, stir until a paste is obtained, and apply the mixture to the stained area. Allow to dry and wash with clean water. Care should be used in its application as it will bleach fabrics and corrode metal items if dropped on these surfaces.

**Oil and Grease**—Petroleum-base materials penetrate concrete readily and are difficult to remove. If the stains are old and deeply imbedded, they are almost impossible to remove. If the stains are fresh and not too deeply imbedded, remove them by covering with an inch or more of dry Portland cement, hydrated lime, or whiting. These materials absorb the stain. Commercial oil-absorbing materials, such as those used by garages and filling stations, can also be purchased.

Another method used to remove this type of stain is by scrubbing with a strong solution of washing soda or trisodium phosphate, using 2 pounds to the gallon of water. Flush with clean water.

**Ink**—Prompt removal by wiping should be attempted. Inks vary in material content, but generally can be removed by one of the following methods:

1. Cover the stain with cotton batting or a heavy cloth bandage saturated with ammonia.
2. Cover the stain with a water-mix paste of chlorinated lime and whiting, allow to dry, remove, and wash.
3. Javelle water may be used in a manner similar to ammonia water. (Javelle water is a strong bleaching solution and not recommended for general cleaning purposes.)
4. Make a strong solution of sodium perborate in hot water, and mix with whiting to form a thick paste. Apply in a 1/4-inch layer over the stain, allow to dry, remove, and wash the surface with water.

**Copper and Bronze Stains**—Mix 1 part of dry sal ammoniac and 4 parts of powdered talc, add water and stir to a thick paste. Place a 1/4-inch layer over the stain. Remove when dry, being careful to use a wooden scraper when working on marble or other fine surfaces.

If the stains are light, dissolve sal ammoniac in household ammonia and apply to the surface with a brush. After about five minutes, flush the surface with clean water.

**Rust**—Minor rust stains can be removed by mopping the surface with a solution of 1 pound of oxalic acid dissolved in 1 gallon of clean water. Apply the solution to the stained area, allow to stand for about 2 hours, and then brush or broom the surface with clean water.

For stubborn rust stains, apply a mixture of 1 part sodium citrate in 6 parts of warm water mixed with 6 parts of commercial glycerin. Mix a part of this solution with enough powdered whiting to form a paste and apply this paste on the affected area. Allow to dry. If the stain has not disappeared, apply a fresh coat of the paste, allow to dry, and wash the area with clean water.

**Smoke and Fire Stains**—Normally this type of stain can be removed by scrubbing with a solution of trisodium phosphate dissolved at a rate of 1/4 pound to 1/2 gallon of water. If the stains are stubborn, follow the method recommended for tobacco stains, washing thoroughly after removing the solution.

**Perspiration Stains**—Perspiration stains from hands can generally be removed by scrubbing with a good commercial grit-type cleaner. If the stain is stubborn, scrubbing the area with a trisodium-phosphate mix of 1/2 lb. to the gallon of water will generally remove the stain. Rinse the area with clean water.

## MISCELLANEOUS DATA

Tables 8, 9, 10, and 11 are included to furnish useful data on earth excavation factors, amounts of concrete required for walls, cubic content of trenches, and weight of standard reinforcing bars.

### Table 8. Earth Excavation Factors

| Depth | Cubic Yards per Square Foot | Depth | Cubic Yards per Square Foot |
|---|---|---|---|
| 2″ | .006 | 4′— 6″ | .167 |
| 4″ | .012 | 5′— 0″ | .185 |
| 6″ | .018 | 5′— 6″ | .204 |
| 8″ | .025 | 6′— 0″ | .222 |
| 10″ | .031 | 6′— 6″ | .241 |
| 1′— 0″ | .037 | 7′— 0″ | .259 |
| 1′— 6″ | .056 | 7′— 6″ | .278 |
| 2′— 0″ | .074 | 8′— 0″ | .296 |
| 2′— 6″ | .093 | 8′— 6″ | .314 |
| 3′— 0″ | .111 | 9′— 0″ | .332 |
| 3′— 6″ | .130 | 9′— 6″ | .350 |
| 4′— 0″ | .148 | 10′— 0″ | .369 |

Example: Assume an excavation 24 ft. x 30 ft. and 6 ft. deep. 24 x 30 = 720. In the table, the 6 ft. depth has a factor of .222 (the number of cu. yd. in an excavation 1 ft. square and 6 ft. deep). 720 x .222 = 159.84 Cu. Yds.

## Table 9. Concrete For Walls
### (Per 100 Square Feet Wall)

| Wall Thickness | Cubic Feet Required | Cubic Yards Required |
|---|---|---|
| 4″ | 33.3 | 1.24 |
| 6″ | 50.0 | 1.85 |
| 8″ | 66.7 | 2.47 |
| 10″ | 83.3 | 3.09 |
| 12″ | 100.0 | 3.70 |

## Table 10. Trench Excavations
### (Cu. Yd. Content Per 100 Lineal Ft.)

| Size (Inches) | Weight (Pounds Per Foot) |
|---|---|
| ¼ round | 0.167 |
| ⅜ round | 0.376 |
| ½ round | 0.668 |
| ½ square | 0.850 |
| ⅝ round | 1.043 |
| ¾ round | 1.502 |
| ⅞ round | 2.044 |
| 1 round | 2.670 |
| 1 square | 3.400 |
| 1⅛ square | 4.303 |
| 1¼ square | 5.313 |

## Table 11. Concrete Reinforcing Bars

| Depth in Inches | Trench Width in Inches | | | | | | |
|---|---|---|---|---|---|---|---|
| | 12 | 18 | 24 | 30 | 36 | 42 | 48 |
| 6 | 1.9 | 2.8 | 3.7 | 4.6 | 5.6 | 6.6 | 7.4 |
| 12 | 3.7 | 5.6 | 7.4 | 9.3 | 11.1 | 13.0 | 14.8 |
| 18 | 5.6 | 8.3 | 11.1 | 13.9 | 16.7 | 19.4 | 22.3 |
| 24 | 7.4 | 11.1 | 14.8 | 18.5 | 22.2 | 26.0 | 29.6 |
| 30 | 9.3 | 13.8 | 18.5 | 23.2 | 27.8 | 32.4 | 37.0 |
| 36 | 11.1 | 16.6 | 22.2 | 27.8 | 33.3 | 38.9 | 44.5 |
| 42 | 13.0 | 19.4 | 25.9 | 32.4 | 38.9 | 45.4 | 52.0 |
| 48 | 14.8 | 22.2 | 29.6 | 37.0 | 44.5 | 52.0 | 59.2 |
| 54 | 16.7 | 25.0 | 33.3 | 41.6 | 50.0 | 58.4 | 66.7 |
| 60 | 18.6 | 27.8 | 37.0 | 46.3 | 55.5 | 64.9 | 74.1 |

# CHAPTER 4

# Carpentry

Because of its availability, wood timbers and lumber have been used to construct almost every type of building and bridge. Wood is divided into two general categories—hardwoods and softwoods. Hardwoods generally come from trees with broad leaves which drop off during winter. Softwoods are those which come from evergreen or needle-bearing trees. These are known as *conifers*. Well-known hardwoods are birch, hickory, maple, oak, walnut, and mahogany. Familiar softwoods are cypress, cedar, redwood, Douglas fir, spruce, and nearly all of the varieties of the pine (except yellow pine). Hardwoods have pleasing grain patterns and can be given a clear finish, but are more difficult to work than softwoods. Softwoods generally resist weathering better than most hardwoods, with cypress, redwood, and cedar particularly recommended for exterior work.

The types of wood also have grade classifications. Lumber free of imperfections falls into one grade, lumber with some imperfections into another, and lumber which has many imperfections falls into a third category. A rule of thumb guide would dictate purchasing select lumber only for interior work where a fine or natural finish is desired, and common lumber for all general construction.

## SEASONING LUMBER

Controlled drying or seasoning of lumber improves its strength. Dead weight of the finished product or construction item is reduced when dry lumber is used, and handling and transportation problems are reduced. Dry lumber makes tighter joints, is easier to size, and is easier to saw and prepare for installation. Because lumber shrinks as it drys, the use of dry lumber prevents warpage and defects in the finished construction.

Seasoning takes time—from a few days for kiln drying, up to many months for air drying under such adverse conditions as cold or humid weather. Lumber in storage should not contact the ground. Therefore, some type of pile foundation is necessary. Foundations often are constructed for piling lumber in open storage, and if these are made, they should slope about 1 inch per foot of length from the front to the rear, with the rear posts high enough to keep the under side of the boards in the first course at least 1 foot above the ground. Lumber is also stored in open sheds, but regardless of how it is stored, it must be kept off the bare ground and have sufficient ventilation to prevent mildew and rot.

## LUMBER SIZES

Lumber is normally available in common or standard sizes of definite widths, thicknesses, and lengths. This permits ease and uniformity in planning structures and ordering materials. The common widths and thicknesses of lumber in use today are listed in Table 1 on page 178. Table 3 on page 179 lists the number of board feet in lumber of typical lengths and cross sections. Common commercial lengths range from 8 to 20 feet, although longer lengths are usually available at extra cost.

## MOLDINGS

There are many types of moldings used in maintenance and carpentry repair and construction. Those shown in Fig. 1 are in most common use.

## DOORS

There are many designs of both exterior and interior panel doors, and these are normally selected to complement the architectural style of the building. Exterior softwood doors are normally

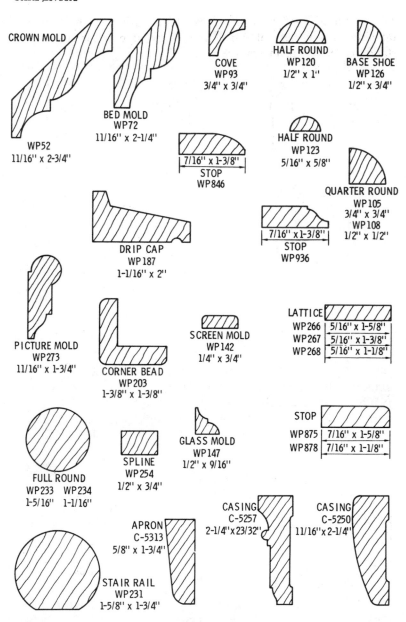

**Fig. 1. Moldings**

1-3/4″ thick, 2′-8″ wide and 6′-8″ high, although they are obtainable in a variety of other sizes. Typical exterior doors with standard details are shown in Fig. 2.

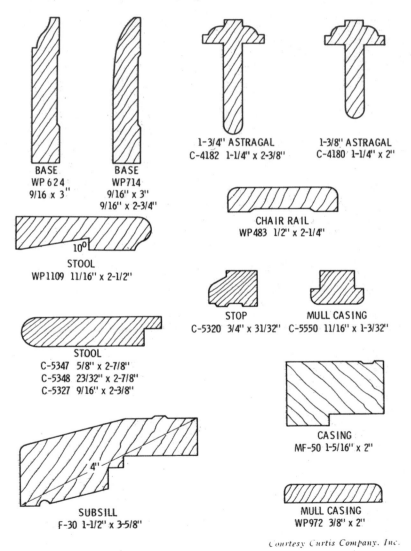

BASE
WP 624
9/16 x 3″

BASE
WP714
9/16″ x 3″
9/16″ x 2-3/4″

STOOL
WP1109 11/16″ x 2-1/2″

STOOL
C-5347  5/8″ x 2-7/8″
C-5348  23/32″ x 2-7/8″
C-5327  9/16″ x 2-3/8″

SUBSILL
F-30 1-1/2″ x 3-5/8″

1-3/4″ ASTRAGAL
C-4182  1-1/4″ x 2-3/8″

1-3/8″ ASTRAGAL
C-4180  1-1/4″ x 2″

CHAIR RAIL
WP483 1/2″ x 2-1/4″

STOP
C-5320  3/4″ x 31/32″

MULL CASING
C-5550  11/16″ x 1-3/32″

CASING
MF-50 1-5/16″ x 2″

MULL CASING
WP972 3/8″ x 2″

*Courtesy Curtis Company. Inc.*

*in common use.*

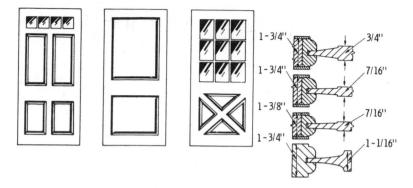

**Fig. 2. Typical exterior panel doors and details.**

Interior doors also vary in design and sizes with the usual thickness being 1-3/8″. Typical types and interior door details are shown in Fig. 3.

Increased use is being made of both the interior and exterior type of hollow-core doors which have an all-wood grid core notched into the stiles for solid light-weight strength. They are faced with veneer (usually 3-ply) glued to the frame and core under pressure to form a strong unit. They are available in all grades and species of stock veneers. Maximum sizes of these doors normally are 4′-0″ × 8′-0″ with a maximum thickness of 2-1/2″. Standard sizes are similar to those of other type doors. Fig. 4 shows some of the designs available in this type of door.

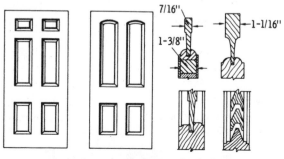

*Courtesy Curtis Company, Inc.*

**Fig. 3. Typical interior panel doors and details.**

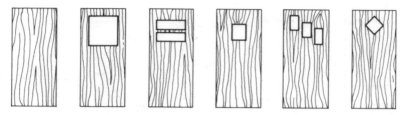

**Fig. 4. Typical designs available in hollow-core doors.**

## Installation of New Doors

Extreme care should be used in placing and installing new wooden doors. The following suggestions are recommended:

1. Doors should be protected in storage and on the job site to prevent extremes of humidity before hanging and finishing. Store on a flat level surface. *Avoid standing doors on edge for storage purposes.*

2. Construction or architectural balance may be ruined if doors are cut down in size to any great degree. Perform only the necessary trimming for fitting. Make every attempt to secure a door for the size of the planned opening so only the minimum amount of trimming will be necessary. The maximum amount any flush door should be trimmed, especially the hollow-core type, is 1/2" on each side.

3. Allow approximately 3/16" clearance between the door and jambs in hanging a door to allow for swelling in extremely damp weather.

4. Use three hinges on all exterior doors. Set the hinges in a straight line so that the door will not warp after installing.

5. The stops and jambs must form a square and the frame must be plumb for the doors to operate properly.

6. Handle unfinished doors with clean gloves. Bare hands may leave finger marks and soil stains which may be difficult to remove.

7. It is good practice to allow at least 1" of wood on all sides of the mortise when mortising for locks.

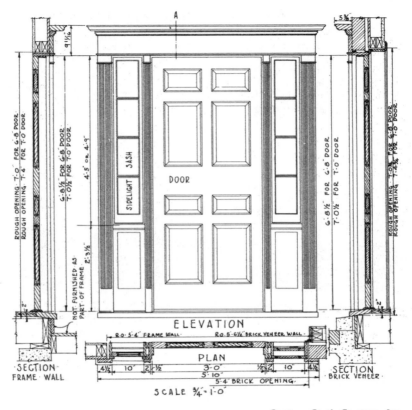

**Fig. 5. Details of a typical residential or small commercial building door.**

8. When finishing the doors, be sure the surface is clean and dry.

9. Wood doors should be dry before applying any finish. Avoid applying the finish during damp or rainy weather, or when the humidity is excessively high.

10. Apply the finish as soon as the door is fitted. Be sure that all four edges, including the top and bottom, receive at least two coats of paint, varnish, or sealer after the door has been fitted and is ready to hang.

11. Handling marks or exposure effects should be removed by sanding lightly with 3/0 or 5/0 sandpaper, cleaning the

surface before applying the finish. Sand again between the finish coats.

12. When applying the finish, follow the finish manufacturer's recommendations very closely.

Fig. 5 shows the detailed elevation of a typical residential or small commercial building door.

## Repair Hints

When removing a door from the frame, it is recommended that the bottom hinge pin be removed first, then the top hinge pin. When the door is replaced, insert the top hinge pin first and then the bottom, making it easier to handle the door.

When planing a door edge, be sure the plane blade is sharp and set for a light cut. Plane from the end toward the middle of the door, holding the plane so the blade meets the wood diagonally. Do not drag the plane back over the edge which could result in splintering the wood.

If an old door sticks, the cause can usually be traced to loose screws in one or more of the hinges. Often, the screw hole has enlarged, and if this has happened, fill the holes with wood putty, let it harden, and then replace the screws. Steel or lead wool can be used in place of putty, with lead wool being preferred. If the door rattles, it is probably the result of a poorly fitted or loosened strike plate. In most cases, properly positioning of the strike plate will correct the trouble.

## Minor Door Repairs

**Door Squeaks** can be quieted by a few drops of oil at the top of each door hinge. Move the door back and forth. If the squeak continues raise the door pin and add more oil.

**Noisy Locks** are lubricated with graphite. Graphite also helps to loosen up tight turning locks.

**To Stop A Rattling Door Knob** loosen the knob set screw and remove the knob. Place a small piece of putty or caulking compound in the knob and put the knob back in place pushing the knob on as far as possible. Tighten the set screw and in almost every instance the rattle in the door stop has been stopped.

**To Cure Dragging or Sticking Doors** tighten the hinge screws. If they do not hold replace with longer screws. Often, placing a toothpick or matchstick in the screw holes and inserting the old screws will permit secure fastening.

For sticking doors open the door slowly and look for a shiny spot on the door. Sand down the spot, closing the door often so as not to remove too much of the door by the sanding process. Too much sanding could create a loose fit of the door.

Occasionally a door frame may warp or go out of shape. Then it may be necessary to remove the door and to plane the door part that is sticking.

## WINDOWS

Windows (sash) are produced in steel, aluminum, and wood, with the double-hung wood type being more commonly used than any other. Details of this type of window are illustrated in Fig. 6.

Sliding windows are preferred by some. Details of the removable sliding type are shown in Fig. 7.

### Stuck Windows

Stuck windows are normally the result of hardened paint in one or more of the grooves of the window sash. As the paint is applied to the sash and frame, it hardens and forms an adhesive seal which is difficult to break without the assistance of some tool. Screwdrivers are not advocated for opening stuck windows. A tool with a wide thin blade, such as a putty knife or scraper, should be used. Insert the flat-bladed tool between the sash and frame, working it from top to bottom to break the paint seal. Take care not to gouge the wood.

If at all possible, the window should be loosened from the outside of the building and at the bottom of the sash. This section is the one most often sealed by the paint. If a putty knife, or other thin wide-blade tool, is inserted under the sash and the paint seal broken, the window should move. If this does not work, then the tool should be carefully placed between the frame and sash at the sides and tapped lightly until the seal is broken. If the window sticks as it moves up and down, locate the section where it binds and remove the paint. If the window sticks at the top, then use the same procedure that was used in freeing the bottom section.

Another procedure used is to place a small block of wood against the sash and tap it lightly with a hammer or mallet. The block of wood should be moved up and down the window sash and the tapping should be done lightly so as not to damage the window or loosen the putty.

If the window is binding because of too much moisture, it may be necessary to remove the entire sash and lightly sand or plane the sticking sections, using care not to remove too much of the sash sides as this may later result in the window rattling if the wood dries.

## FLOORS

Experimentation has proved that minor floor squeaks can sometimes be eliminated by the application of a generous supply of talcum powder to the floor board areas where the squeaks occur. Wipe the powder into the cracks with a cloth and then wipe the floor clean. This method, of course, will not work in all cases.

If the floors can be seen from underneath, as in a basement or crawl space, locate the squeak areas from the underside of the floor. When this is done, drive wooden wedges into place between the underside of the floor and the joists at the squeak points. If this procedure does not solve the problem, cut a piece of $2'' \times 4''$ to fit between two joists, forcing the piece of wood up against the floor. Drive nails through the joists into the piece of $2'' \times 4''$ so that it is held tightly in place.

For stubborn areas, it may be necessary to drive screws up through the subfloor and part way into the finish floor to bring the two floors together. Care must be used to select the right length of screw so that their points will not go completely through the finish floor, but still long enough to hold tightly.

If the floor is such that it cannot be reached from the underside, locate the joists in the trouble area. At the squeak point, drive a nail (preferably a threaded or screw-type nail, as they have more holding power) at a slight angle through the floor and into the joist. At a point about an inch away, drive a second nail at an angle below the floor surface and fill the holes with plastic wood. Sand the surface after the plastic wood has dried. Finish the areas to match the existing floor finish.

## WOODEN STAIRS

The creaking in wooden stairs can often be eliminated by bringing the treads and risers together if they have become separated. Refastening is normally done with nails if the treads and risers are the butt-joint type. If the treads and risers are assembled with groove-and-rabbet joints, wooden wedges can be used.

DOUBLE-HUNG WINDOWS 2 X 9

COMBINATION UNITS
WITH PICTURE SASH

VERTICAL SECTION

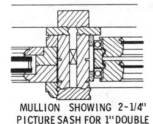

MULLION SHOWING 2-1/4"
PICTURE SASH FOR 1" DOUBLE
INSULATED GLASS GLAZING

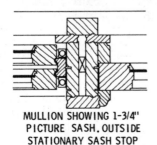

MULLION SHOWING 1-3/4"
PICTURE SASH, OUTSIDE
STATIONARY SASH STOP

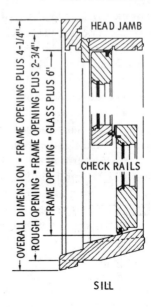

*Fig. 6. Details of a*

## NAILS

The most common method of fastening pieces of wood or timber together is with nails. Although many varieties are being sold, those most in use are the common, box, casing, cut, finishing. and brad types. The *common nail* is flat headed, has a diamond-

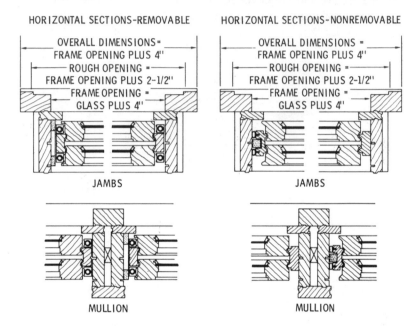

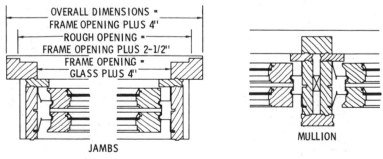

*Courtesy Curtis Company, Inc.*

**double-hung window unit.**

shaped point, and is generally used for rough work where strength is an important factor. The *box nail* is similar to the common nail, but is thinner and used where the heavier-type common nail might split the wood.

*Casing nails*, *brads*, and *finishing nails* are generally set or driven below the surface of the wood. The heads are only slightly larger than the shank.

*Cut nails* have flat sides, have better holding power than regular nails, and are generally used in nailing flooring. Hardened cut nails are available, and are often used as masonry fasteners.

HORIZONTAL SECTION SLIDING WINDOW UNIT

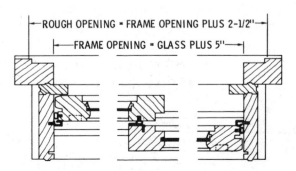

|←—ROUGH OPENING = FRAME OPENING PLUS 2-1/2"—→|

|←—FRAME OPENING = GLASS PLUS 5"—→|

*Fig. 7. Sliding*

When driving nails, blows that are too heavy often cause the nail to tear the wood fibers, decreasing their holding power. Drive with only enough power to force the nail into the wood. When driving nails into hardwood, a recommended practice is to first drill an undersized pilot hole into the wood. The use of paraffin or beeswax also eases the difficulty of driving nails into hard-

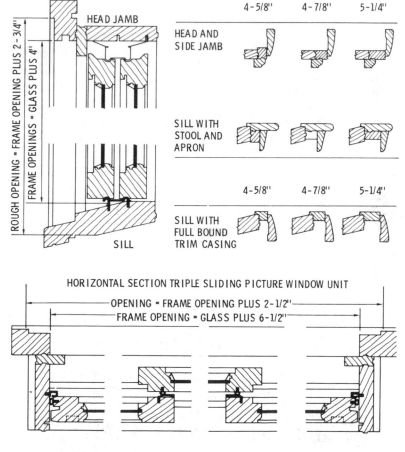

Courtesy Curtis Company, Inc.

**window details.**

wood. Nails hold better when driven into the cross grain of the wood.

*Threaded nails* are now in general use throughout the construction industry. Tests conducted by some of the manufacturers indicate that this type of nail is less likely to split certain types of wood, has better holding power, and will withstand a great degree of contraction and expansion which often loosen common or regular nails.

## WOOD SCREWS

Although the most commonly used method of fastening is with nails, the wood screw also is in everyday use. It has been estimated that a wood screw has more than twice the holding power of a nail when used in softwoods and approximately four times the grip when used in hardwoods.

Softwoods normally take screws easily and require only a prick point with an awl or other pointed tool to start the screw. However, in hardwoods such as oak, walnut, maple, or birch, a pilot hole should be drilled so as to prevent the possibility of splitting the wood.

There are many types of wood screws, the most common being the flathead type where countersinking of the application point is necessary so that the head rests below the finish surface. These countersunk areas can be filled with plastic wood or wood putty and sanded down to match the top of the surrounding wood surface.

Other types of wood screws include round head, phillips, and the lag screw with the square head used generally for rugged work.

Wood screws are made in a variety of metal types, the most common being those made of steel. Steel screws used outdoors should be dipped into a paste mix made of powdered graphite and linseed oil prior to being driven into place. This procedure may prevent rusting. However, for outdoor and special work, brass or rust-resisting screws are recommended.

It is essential that a proper screwdriver be used to drive screws into place. Screw slots and wood surfaces are often damaged because of using the wrong screwdriver. The screwdriver blade tip should fit the screw slot properly, being neither wider nor narrower than the slot in the screw head.

## SANDPAPER

Although sandpaper can be purchased by number, the most common practice is to purchase it by type, such as fine, medium. coarse, and very coarse. Occasionally, a manufacturer will recommend that sanding discs or certain types of sanding belts be purchased for his equipment by number. In this case it is best to follow the manufacturer's recommendations.

There is little or no sand used in sandpaper manufacture. The mineral generally used is flint (white quartz). Flint paper is the most inexpensive type and should be used only for light work. For heavier work, garnet paper (made of red quartz), is often used by carpenters and cabinet makers. It is hard, and has good abrasive power. Emery paper is used in metal work. Aluminum-oxide and silicon abrasives have better than average durability and are used on both metal and wood. Both types are long lasting and are used extensively with power sanding tools. Basically, coarse abrasive paper is used for heavy-duty work, medium for a smoother finish, and fine to prepare wood surfaces for a seal or other finishing material.

When sanding by hand, back the paper with a wood or a metal block. If the sanding is the final preparation before applying a finish, it is good practice to use a backing block with some "give," such as felt or rubber material. Blackboard erasers make excellent backing blocks for fine sanding of wood. Good hand-sanding results are obtained by using a metal-block backing for coarse work, a wood block for medium sanding, and felt blocking for fine work.

When purchasing sandpaper for use with power sanders, it is best to buy the type of paper recommended by the power-tool manufacturer for the particular machine involved. This will result in the most satisfactory performance.

## GYPSUM WALLBOARD

A construction material that is used extensively in the industry for building use is gypsum wallboard. This prefabricated plaster, which has a tough surface cover of paper on both sides, is easy to cut and install.

In preparation for installation, the board is generally placed over a bench, sawhorses, or other flat surface, and marked lightly for cutting if necessary. A knife is drawn along the marked line, cutting through the surface paper. The cut is then held over the edge of the surface it is on and the board is given a quick downward snap, causing it to break cleanly along the scored line. The paper on the back side is then cut along the crease formed by this action.

Perforated tape and thin coats of prepared cement are placed over the joints and thin coats of prepared cement to the joint area surfaces complete the job. The joints are sanded after they dry to prepare the wall for painting.

### Patching Holes in Plaster and Wallboard

There are two common types of material used for patching holes in plaster and wallboard namely—spackling compound and patching plaster.

Spackling compound is convenient and should be used for small repair jobs as it can be purchased in small amounts. Patching plaster is generally packaged in larger packages and is generally less costly than the spackling compound. Both can be used to advantage and both must be mixed with water to a paste consistency before use.

**To patch a small hole or crack:**

a. Remove any loose plaster with a knife and remove plaster from the back edges of the crack so as to make the back edges of the crack or hole wider than the front.

b. Totally dampen the opening surface with water using a sponge, cloth or paint brush.

c. Mix a small amount of the compound to a paste consistency.

d. Fill the opening with the patching material pressing the mixture until it completely fills the hole. Smooth the surface with a putty knife. After it dries wrap a piece of sandpaper around a small wood block and sand the surface. The wooden block provides an even sanding surface.

e.  For larger holes or cracks use a two-step operation. After preparing the opening of the hole or crack partly fill the opening and allow the patch to dry. Add a second paste mix of the compound and allow this to dry. After the second coat dries sand to a smooth finish.

f.  For large holes it may be necessary to fill in area with wadded newspaper. Start patching by working the patching mix from all sides. Allow this mix to dry and then apply another layer around the new edge continuing this process until the hole is filled. After patch work has been completed and material is thoroughly dry, sand to a smooth finish.

Note: For a textured plaster surface use a sponge or comb to create a surface similar to that existing before the patch is dry.

Perforated tape and thin coats of prepared cement are placed over the joints and thin coats of prepared cement applied to the joint area surfaces complete the job. The joints are sanded after they dry to prepare the wall for painting.

## PLYWOOD PANELING

*(Information and illustrations courtesy Georgia-Pacific Corp.)*

Plywood paneling is available in nearly any kind of wood-grain surface, either sanded ready to finish or already prefinished. The usual thickness of panels used for most wall covering is $1/4''$, although other thicknesses are available. The standard size sheets are $4' \times 8'$; though other lengths are available on special order. Plywood paneling is also available in random-plank sizes.

### Preparation

If remodeling, remove all existing trim. Plan your work and sketch the job before you start. Doing this will save considerable time and materials when construction begins.

Stand the panels around the walls and position them to obtain the most pleasing effect of color and grain. Number the panels in sequence to avoid confusion. Accurate measuring, sawing, and fitting of the panels are necessary to obtain a perfect installation.

## Installation

The following procedures are recommended for installing plywood paneling.

**Existing Plaster or Other Nonmasonry Walls**—Use 3/8″ × 1-7/8″ plywood strips or 1″ × 2″ lumber for furring. If the walls are flat and true, the panels may be nailed or glued directly to the walls. If nails are used, nail through the panels into the studs.

**Existing Masonry Walls**—Use 1″ × 2″ lumber furring, or 2″ × 3″ or 2″ × 4″ framing laid flat against the wall.

**New Construction**—Use 5/16″ plywood sheathing or 3/8″ gypsum board for underlayment, or 3/8″ × 1-7/8″ plywood furring strips.

## Furring Tips

The following suggestions will be found useful when furring a wall for the application of paneling.

1. The furring strips should be applied so that the face grain runs at right angles to the direction of the grain in the panels.
2. The furring strips should be installed horizontally and spaced 16″ on center, with filler strips spaced 48″ on center, as shown in Fig. 8.

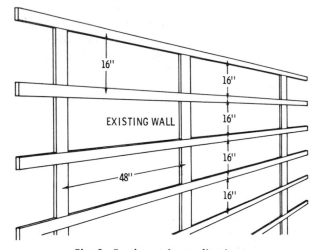

*Fig. 8. Furring strip application.*

3. Nail through the paneling and furring into the studs for solid construction.

4. When fastening into masonry, drill holes, insert wooden plugs or expansion shields, then nail or screw the furring to the plugs or shields. Other types of fasteners, such as nail adhesive anchors, concrete nails, or drive pins can be used by following the manufacturer's directions.

5. All four edges of each panel must be attached to the furring. Add strips as necessary.

6. The furring strips or framing should be shimmed to provide a true flat wall.

### Perfect Panel Joints

To obtain close-fitting joints, scribe the panels for all corners and around doors, windows, fireplaces, etc. To achieve a perfect fit in corners that are out of plumb, and around brick, stone, etc., position the panel for the proper height, either on the floor or on a level block. With a compass, and using the corner or the brick or stone surface as a guide, scribe a line from the top to the bottom of the panel, as shown in Fig. 9. Cut along this line with a coping saw. Where the panels join each other, bevel the edge slightly toward the back of the panel with a small block plane to provide a tighter fitting joint.

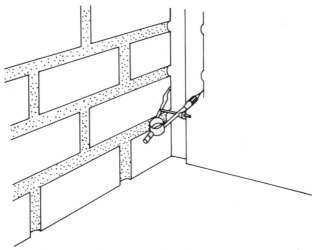

*Fig. 9. Scribing a panel to fit in a corner.*

135

## Starting

Begin at one corner of the room, and scribe and plumb the first panel to fit the corner. Use a compass to scribe the panel, as shown in Fig. 10. Nail the panel into place. Butt the second panel against the first and nail into place.

*Fig. 10. Scribing the starting panel.*

*Fig. 11. Method of nailing panel-ing direct to wall studs.*

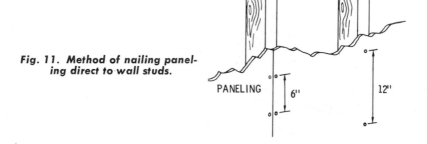

For a large area of paneling, allow a small space at the end of the wall for expansion. Also allow approximately 1/4″ clearance at the top or bottom. Molding will cover this expansion space. Changes in humidity and temperature within the room will cause the panels to swell and shrink slightly. If they are butted tightly against the floor and ceiling, they may buckle and pull loose from the wall.

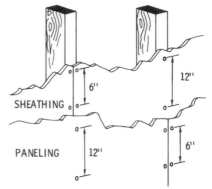

**Fig. 12. Method of nailing paneling over underlayment or old walls.**

## Fastening Methods

**Nailing**—Whether application is direct to the studs, over an underlayment, or over an old wall, the panel edges should join over a stud. For application direct to the studs, use 3-penny (1-1/4″) finishing nails spaced 6″ along the panel edges and 12″

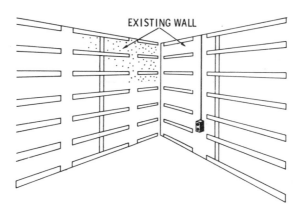

**Fig. 13. Horizontal and vertical furring strips applied to a masonry wall.**

elsewhere. These dimensions are shown in Fig. 14. For applications over furring, use 3-penny (1-1/4″) finishing nails spaced 8″ along the edges of the panel and 16″ (into the horizontal furring) elsewhere. By nailing in the panel grooves, the nail heads are easily hidden. Set the nail heads 1/32″ deep.

137

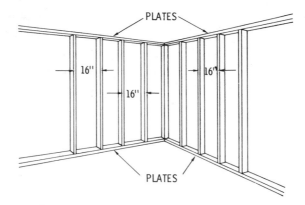

**Fig. 14. Vertical studs erected on the face of a masonry wall.**

For application over underlayment or old walls, use 6-penny (2″) finishing nails spaced 4″ along the panel edges and 6″ elsewhere (Fig. 15). Set the nail heads 1/32″ deep and fill the holes with a putty stick colored to match the paneling.

**Gluing**—Contact cement may be used in place of nails to fasten the panels to the wall. Be sure to cut and fit each panel exactly before the contact cement is applied because, once applied, it is impossible to make adjustments. *Note: This method is not recommended for use over plaster walls in poor condition.*

Position the first panel (in the corner) so that it contacts the corner edge first, then swing the panel into place. Follow the same procedure for all the other panels, butting the edge of the panel to be installed against the previously installed panel, and then swinging it into place.

### Paneling Over Masonry Walls

For masonry walls having one side facing the outside of the house, either above or below grade, be sure the wall is properly waterproofed before the studding, furring, or paneling is applied. Where excessive humidity may cause condensation on the interior surface of exterior masonry walls, use a vapor barrier paper or film over the furring to eliminate moisture condensation and to prevent any that might occur from penetrating to the paneling.

If the masonry walls are true and flat, use masonry nails to apply horizontal 1" × 2" furring strips every 16" beginning 1/4" above the floor level. Insert vertical furring strips to support the edges of the panels (Fig. 13). Shim as necessary to make the surface of the furring true and flat. Allow a breather space of at least 1/4" above the top and bottom furring strips. Breather spaces should also be provided between each vertical furring strip and the horizontal furring.

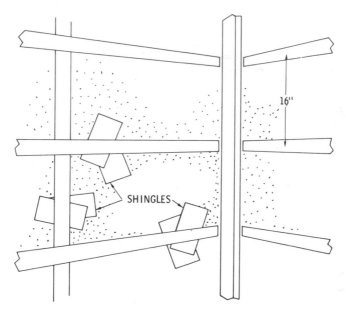

**Fig. 15. Shims placed behind furring strips
to level any uneven surfaces.**

If the masonry walls are uneven, it may be necessary to install 2" × 3" or 2" × 4" studs as shown in Fig. 14. Use top and bottom plates and span the studs 16" on center. Apply the panels directly to the studs or to an underlayment as previously explained.

## Paneling Over Old Plastered Walls

Paneling may be applied directly over old plaster if the wall is true and not badly cracked. If the walls are in poor condition, however, or if they are uneven or heavily textured, furring strips

139

should be used. Nail $1'' \times 2''$ strips horizontally to the studs, starting at the floor level and continuing up the wall at 16″ inter-

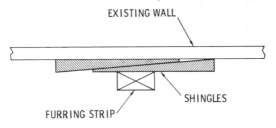

Fig. 16. Wood shingles applied in this manner make ideal shims.

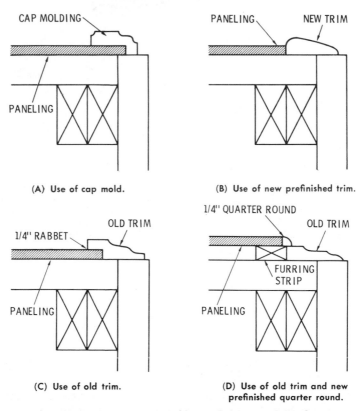

(A) Use of cap mold.

(B) Use of new prefinished trim.

(C) Use of old trim.

(D) Use of old trim and new prefinished quarter round.

Fig. 17. Four ways to panel around doors and windows when remodeling.

vals. Nail 1″ × 2″ vertical strips every four feet between the horizontal strip (leaving a breathing space) to support the panel edges. This method is illustrated in Fig. 15. Place shims behind the furring strips to level any uneven areas. Wood shingles make ideal shims if applied in the manner shown in Fig. 16. Short pieces of bevel siding can also be used.

## Paneling Around Doors and Windows

**Old Construction**—Special care is needed to provide a neat appearing junction of the wall paneling and the existing trim around windows and doors. There are several methods of accomplishing this without the necessity of scribing and sawing the panel to butt tightly against the trim.

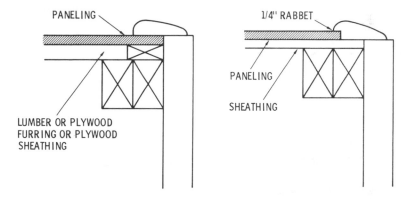

(A) Use of prefinished trim installed flush over paneling and jamb.

(B) Use of a rabbeted prefinished trim where paneling is installed over sheathing.

*Fig. 18. Two ways to panel around doors and windows in new construction.*

One method (Fig. 17A) is the use of a cap molding to cover the edge of the paneling at the door or window opening. Make sure the cap molding is set back slightly from the edge of door jambs to allow space for the hinges.

The method shown in Fig. 17B uses new prefinished trim butted tightly against the edge of the paneling. Where it is desirable to use the old trim, the method shown in Fig. 17C works quite

141

well. Remove the trim and rabbet a 1/4″ recess in the back side, as shown. Replace the trim and insert the paneling.

Where furring strips have been used, apply a 1/4″ prefinished quarter round, as in Fig. 17D, to hide the panel edge.

**New Construction**—Two methods of paneling around doors and windows in new construction are shown in Fig. 18. In Fig. 18A,

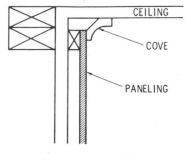
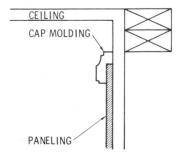

(A) Use of cove mold at the ceiling joint.

(B) Use of a cap molding where paneling does not reach the ceiling.

**Fig. 19. Two ways to treat the paneling at the ceiling.**

a prefinished trim is applied over the joint between the paneling and the jamb to cover the crack. In Fig. 18B, the paneling is applied over sheathing or furring strips, in which case the molding is rabbeted as shown to receive the panel.

## Treatment of Ceiling Joints

The joint between the paneling and the ceiling can be concealed by using a cove mold (Fig. 19A), or with a cap molding (Fig. 19B) where the paneling does not extend to the ceiling line.

## Inside and Outside Corners

Inside corners can be finished in any one of the three ways shown in Fig. 20. The butt joint is prepared by the scribed method explained previously. A small cove mold can be used to cover the corner joint, or an inside corner mold can be inserted if the panel edges are scribed to fit properly.

Outside corners can also be finished in any one of three ways shown in Fig. 21. Where the panel edges butt against the corner molding, and where the panel edges are mitered and butted, re-

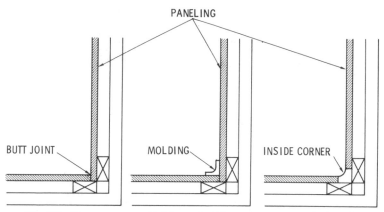

*Fig. 20. Three ways to finish an inside corner on paneled walls.*

quires accurate scribing and fitting. Where this extra care is not desirable, the use of an outside corner molding applied over the paneling at the corner provides a neat and satisfactory corner treatment.

## Panel Treatment At Floor Level

When remodeling, carefully remove the old base molding. Fasten the paneling to the furring strips and reapply the old base molding. Use shoe molding, as in Fig. 22, to complete the job.

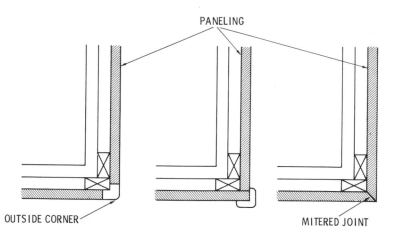

*Fig. 21. Three ways to finish an outside corner on paneled walls.*

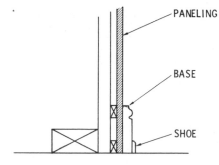

Fig. 22. *Use of old base molding when remodeling.*

If a new base molding is to be used, remove the old shoe molding, install the paneling, and apply new base and shoe molding, as in Fig. 23.

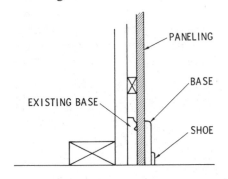

Fig. 23. *Use of new base and shoe molding.*

## FIR PLYWOOD

*(Information and illustrations courtesy American Plywood Assn.)*

The plywood most familiar to the carpenter is the standard sheets of fir plywood that can be purchased from any lumberyard. These sheets are 4′ × 8′ and are available in standard thicknesses of 1/4″, 3/8″, 1/2″, 5/8″, and 3/4″. Other sizes and thickness are available for special purposes, but usually are not stocked by the dealer.

Plywood comes in two types—*interior* and *exterior*—and in many grades. Grade AD, for example, indicates that one side contains no knots or imperfections that would mar its appearance if finished natural. The D indicates the back side contains open knots and imperfections impossible to hide with paint. This type of plywood would be used where only one side is to be exposed to view. Grade AA would be used where both sides of the plywood are to be visible.

Fir plywood is also available with unsanded surfaces. This type is produced for use as sheathing and for subflooring where the surface appearance is unimportant. Unsanded plywood can also be used as an underlayment in some applications.

Exterior plywood is made with 100% waterproof glue for outdoor use or where the plywood will be exposed to moisture indoors. Use the interior type for projects such as cabinets and furniture. Within each type are several appearance grades (AA, AC, AD, CD, etc.), so be sure to select the grade best suited for your particular requirements.

### Working With Plywood

Fir plywood is extensively used for large and small projects in industry as well as in the home. The large size of the plywood sheets simplifies every step of construction. It is only necessary

**Fig. 24. Keeping the saw at a low angle reduces splitting of the underside.**

to lay out the work for cutting before preceding with the actual construction.

The lay out should be done with care in order to avoid waste and to simplify the work. When it is necessary to cut several pieces from one plywood sheet it is best to sketch the arrangement of the proposed cuts on a piece of paper before marking the plywood for cutting. Be sure to allow for the saw kerf between adjacent pieces. Try to arrange the work so that the first few cuts will reduce the plywood panel to pieces small enough for easy handling.

**Fig. 25. When using an electric hand saw, the plywood panel is sawed from the back.**

One of the most important points to watch when planning the sequence of operations is to cut all mating or matching parts with the same saw setting. Also watch the direction of the face grain when cutting. Cuts should be made along the grain where possible. Mark the side of the panel having the best surface for cuts to be made with a hand saw, radial saw, or bench saw. Mark the back of the panel and saw from that side when using a portable electric saw.

*Fig. 26. Always plane from the end toward the center when planing the edge of a piece of plywood.*

**Hand Sawing**—When hand-sawing, place the plywood with the good face up, and support the panel firmly so it won't sag. Use a saw having 10 to 15 points to the inch. Splitting out of the underside can be reduced by putting a piece of scrap lumber under the panel and sawing it along with the plywood. It also helps to hold the saw at a low angle, as shown in Fig. 24. Most important of all, *use a sharp saw*.

**Portable Power Saw**—When using a portable power saw, turn the good face of the plywood down (Fig. 25). Tack a strip of scrap lumber to the top of each sawhorse to prevent damage to the top of the horse. Keep the saw blade sharp, as a dull blade will cause the plywood to splinter.

**Planing Plywood Edges**—Planing plywood edges with a plane or jointer won't often be necessary if the cuts are made with a sharp saw blade. If any planing is done, work from both ends of the edge toward the center, as in Fig. 26, to avoid tearing out the interior plies at the end of the cut. Use a plane with a sharp blade and take very shallow cuts.

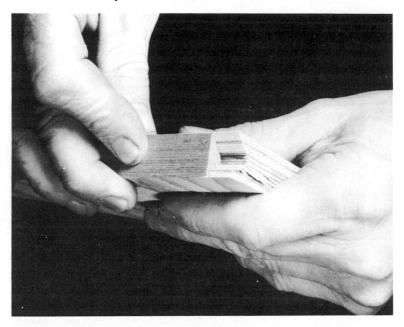

*Fig. 27. Use of a V groove and V-shaped strip to hide a plywood edge.*

## Edge Treatment

Some sort of treatment is usually necessary to either conceal or improve the appearance of plywood edges. The laminated cross section that is revealed when viewing the plywood edge is not always attractive nor does it always blend well with the rest of the unit.

Three ways to finish plywood edges are described here. As shown in Fig. 27, a handsome solid effect can be obtained by cutting a V groove in the edge of the plywood and inserting a

V-shaped solid wood strip. This method is comparatively difficult, however, and nearly impossible except with power tools.

A much simpler method is by applying thin strips of real wood now available that has the back coated with a pressure-sensitive adhesive. To apply, simply peel off the backing paper and press the wood strip to the plywood edge (Fig. 28) according to the manufacturer's recommendations. Edge banding is also available without an adhesive coating. Contact cement is used to bond this type of real-wood banding to the plywood edge.

Laminated plastic surfacing materials may be applied to the plywood edges with the same cement used to apply this material

**Fig. 28. A tape of real wood used to hide a plywood edge.**

to table and counter tops. As shown in Fig. 29, apply the edges first, then the counter or table top. A thicker and more massive effect can be obtained by nailing a 1″ or 1-1/4″ strip underneath the edge.

**Fig. 29. Laminated plastic to match the table or counter top can be applied to the plywood edges.**

Plywood edges that are to be painted should be filled. Several varieties of wood putty are available for this purpose, either powdered to be mixed with water, or already prepared and ready for use. Plaster spackling compound also works well. Whatever type used, allow it to dry thoroughly, sand smooth, and apply the finish.

### Plywood Fasteners

The size of the nails to use in plywood construction is determined primarily by the thickness of the plywood. Used with glue, all the nails shown in Fig. 30 will produce strong joints. For 3/4″ plywood, use 6d casing nails or 6d finishing nails; for 5/8″ plywood, use 6d or 8d finishing nails; for 1/2″ plywood, use 4d or 6d; for 3/8″ plywood, 3d or 4d; for 1/4″ plywood, use 3/4″ or 1″ brads, 3d finishing nails, or (for backs where there is no objection to the heads showing) 1″ blued lath nails. Substitute casing for finishing nails wherever a heavier nail is required.

Predrilling is occasionally called for in careful work where the nails must be placed very close to an edge. As shown in Fig. 31, the bit should be slightly smaller in diameter than the nail to be used. Space the nails about 6″ apart for most work (Fig. 32). Closer spacing is necessary only with thin plywood where there may be a slight buckling between nails. Nails and glue work together to produce a strong, durable joint.

Flat-head wood screws (Fig. 33) are useful where nails will not provide adequate holding power. Glue should also be used if possible. The sizes shown in Table 3 on page 180 are minimums; use longer screws when the work permits. This table gives the plywood thickness, the diameter and length of the smallest screws recommended, and the size of the hole to drill.

Screws and nails should be countersunk and the holes filled with wood dough or putty, as shown in Fig. 34. Apply the putty

**Fig. 30. Relative sizes of nails to use for the different thicknesses of plywood to construct strong joints.**

Fig. 31. Predrilling is sometimes necessary.

Fig. 32. Nails should be spaced about 6 inches along the plywood edge.

so it is slightly higher than the plywood, then sand level when dry. Lubricate the screws with soap if they are hard to drive. Avoid damage to the plywood surface by using Phillips-head screws.

*Fig. 33. Relative sizes of flat-head screws used with different thicknesses of plywood.*

Corrugated fasteners (Fig. 35) can reinforce miter joints in 3/4″ plywood and hold them together while the glue sets. For some kinds of plywood jobs, sheet-metal screws are valuable; they have more holding power than wood screws but come only in short lengths and do not have flat heads. Bolts and washers are good for fastening sectional units together and for installing legs, hinges, or other hardware when great strength is required.

### Gluing Plywood

Choose the glue to use from Table 4 on page 180. Before applying the glue, make sure of a good fit by testing the joint, as

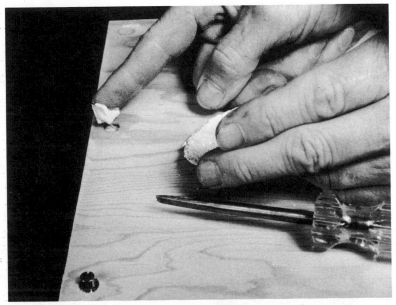

*Fig. 34. Screw and nail holes should be countersunk and filled with wood dough or putty.*

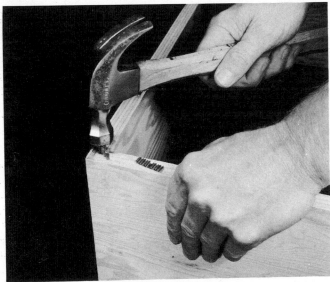

*Fig. 35. Corrugated fasteners can be used to strengthen mitered joint.*

shown in Fig. 36. For lasting strength, both pieces should make contact at all points.

Apply the glue with a brush or stick, in the manner shown in Fig. 37. End grain absorbs glue so quickly that it is best to apply a preliminary coat, allow it to soak in for a few minutes, then apply another coat before joining the parts.

*Fig. 36. Testing a plywood joint before gluing.*

Clamp the joints tightly with clamps, as shown in Fig. 38, or with nails, screws, or other fasteners. Use blocks of wood under the jaws of the clamps to avoid damage to the plywood. Wipe off all excess glue, since some glues will stain the wood and make it difficult to achieve a good finish. Test for squareness, then allow the glue to set.

## Assembling Plywood

Planning pays off in assembly steps, just as in cutting parts. Frequently, the easiest solution is to break down complicated

155

Fig. 37. Applying glue to the surface of a plywood joint.

Fig. 38. Proper clamping is necessary for a strong plywood joint.

projects into subassemblies. These are simpler to handle and make joints more accessible, as shown by the partitioned shelves in Fig. 39. Apply clamps with the full jaw length in contact. When the jaws are not parallel, as at the right in the illustration, pressure is applied to only part of the joint.

A handy, little-known trick for clamping miter joints in cabinets is shown in Fig. 40. With paper sandwiched between to permit easy removal, glue triangular blocks to the ends of each mitered piece. Let the glue set. Apply glue to the mitered ends and pull them together with the clamps. Remove the clamps after the glue has set, pry the blocks away and sand off the paper.

Special clamps frequently save work and result in a better job. Fig. 41 shows various types of edge-clamps used to glue wood

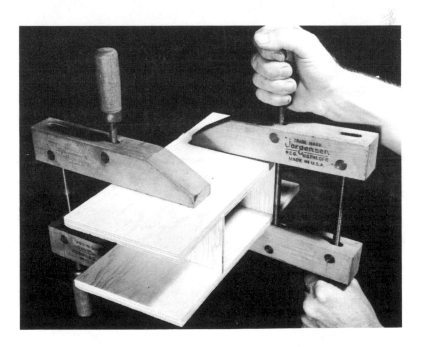

*Fig. 39. Proper and improper application of clamps.*

or plastic edging to plywood. Bar clamps or quick C-clamps grip the panel which is protected by scrap wood. Then the edge-

*Fig. 40. Triangular blocks used to clamp a mitered joint.*

clamping fixtures are inserted to bear against the edge-banding material while the glue sets.

### Installing Plywood

Frame walls permit hanging cabinets by the use of long wood screws through the cabinet backs. The screws must be driven into the wall studs to secure good holding power. Locate the first stud by tapping the wall, then measure off 16″ intervals to find the other studs.

Hollow masonry walls call for the use of toggle bolts or *molly* fasteners. First drill a hole in the masonry with a star drill or carbide-tipped bit, then insert the *molly* and tighten. The bolt can now be removed and used to hang the cabinet.

Concrete, stone, or other solid masonry walls call for anchor bolts. Fasten the base of the anchor to the wall with black mastic, letting it squeeze through the holes. Hang the plywood unit after the mastic has set, using washers. Bolts in expansion shields also may be used.

**Fig. 41. Special clamps used to glue edging to plywood.**

## Drawer Construction

The drawer shown upside down in Fig. 42 is easily made with only a saw and hammer. The butt joints are glued and nailed. The bottom should be 3/8" or 1/2" plywood for rigidity. The drawer front extends down to cover the front edge of the bottom panel.

An additional strip of wood glued and nailed to the front panel, as in Fig. 43, reinforces the bottom of this second type of drawer made with hand tools. This reinforcing permits the use of economical 1/4" plywood for the drawer bottoms.

Power tools make sturdy drawers easy to build. Fig. 44 shows one side of a drawer (dadoed on the outer face for a drawer

Fig. 42. A simple drawer requiring only a hammer
and saw to construct.

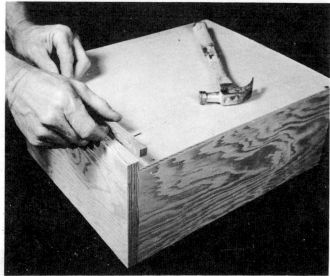

Fig. 43. A strip of wood is used to reinforce this simple drawer.

**Fig. 44. Dadoes made with power tools makes drawer construction easy.**

guide) being put into place. Rabbet the drawer front (at right) to take the sides; dado the sides to fit the drawer back. All four parts are grooved to take a 1/4″ plywood bottom.

A simple drawer guide calling for use of power tools, is shown in Fig. 45. The cabinet side has been dadoed before assembly and a matching strip has been glued to the side of the drawer. Even heavy drawers slide easily on guides like these if waxed or lubricated with paraffin after finishing.

**Repairing Drawers**

If the handles or knobs are loose, tighten the screws from the inside of the drawer with a screw driver. If the knobs are missing or badly damaged new knobs are required. Thread spools have been used as replacements for lost or damaged door knobs as have short lengths of broomsticks with drilled holes.

161

### Sticking Drawers

Remove the drawers and sand down shiny spots until drawer moves easily in its place. Rub the drawer and the frame in those spots where they touch with candle wax or soap. This helps the drawer to glide easier.

If the drawers stick only in damp weather wait until the weather is dry and coat the unfinished surfaces with wax or a penetrating wood sealer.

*Fig. 45. One type of drawer guide made with power tools.*

## HANGING SHELVES

The neatest and strongest way to hang a shelf is by making a dado joint or by using metal shelf supports. A dado requires power tools and does not permit changing the shelf height. Inexpensive shelf supports that plug into blind holes 5/8″ deep drilled in the plywood sides of the cabinet are shown in Fig. 46. Drill additional holes to permit moving the shelves when desired. Another method is the use of slotted metal shelf strips into which shelf supports may be plugged at any height. For a better fit, set the shelf strips flush in a dado cut, or cut out the shelves around the shelf strips.

**Fig. 46. Metal shelf supports that are inserted in blind holes drilled in the sides of a cabinet.**

## SLIDING DOORS

Close-fitting plywood sliding doors are made by rabbeting the top and bottom edges of each door, as in Fig. 47. Rabbet the back of the front door, and the front of the back door. This lets the doors almost touch, leaving only a little gap and increasing the effective depth of the cabinet. For 3/8″ plywood doors rabbeted half their thickness, plow two grooves in the top and bottom of the cabinet 1/2″ apart. With all plywood doors, seal all the edges and give the backs the same paint treatment as the front to maintain the balanced construction of the plywood.

For removable doors (Fig. 48), plow bottom grooves 3/16″ deep and the top grooves 3/8″ deep. After finishing, insert the door by pushing it up into the excess space in the top groove,

163

Fig. 47. By-passing sliding plywood doors with the top and bottom edges rabbeted.

Fig. 48. Top grooves plowed twice as deep as the bottom grooves permit removal of these sliding doors after assembly.

then drop it into the bottom groove. Plowing can be eliminated by the use of a fiber track made for sliding doors of this type.

Only hand tools are required for sliding doors like those in Fig. 49. The front and back strips are stock 1/4" quarter-round molding. The strip between is 1/4" square. Use glue and brads or finishing nails to fasten the strips securely.

**Fig. 49. Only hand tools are needed to construct this type of sliding door.**

## CABINET BACKS

The standard method of applying backs to cabinets and other storage units calls for rabbeting the sides. The cabinet at the left in Fig. 50 has a rabbet just deep enough to take the plywood back. For large units that must fit against walls that may not be perfectly smooth or plumb, the version at the right in Fig. 50 is better. This rabbet is made 1/2" or even 3/4" deep. The lip that remains after the back has been inserted may be easily trimmed wherever necessary to get a good fit between the plywood unit and the wall. When hand tools are used, attach strips

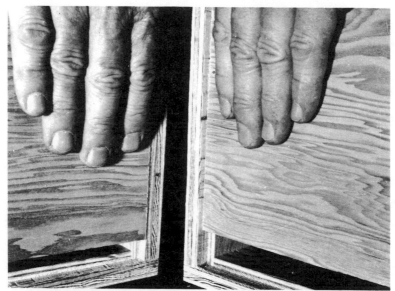

Fig. 50. Two methods used to install plywood backs on cabinets.

Fig. 51. Quarter round can be used to provide a fastening
surface for a plywood back.

of 1/4″ quarter-round molding for the back to rest against (Fig. 51). Glue and nail the back to the molding.

*Fig. 52. Two alternate methods of installing a plywood back where appearance is not important.*

Fig. 52 shows two methods of applying cabinet backs without rabbets or moldings. One method is to nail the back flush with the outside edge. The second method is to set the back 1/2″ to 7/8″ away from the edges. The back becomes inconspicuous when the cabinet is against the wall.

Bevel the cabinet backs that must be applied without a rabbet to make them less conspicuous. Install a 3/8″ plywood back flush with the edges of the cabinet, then bevel with light strokes of a block plane, as shown in Fig. 53.

Nail the cabinet back into the rabbet by driving nails at a slight angle, as indicated in Fig. 54. Use 1″ brads or 4d finishing nails. Where the back will not be seen, the 1″ blued lath nails shown here may be used.

A two-hand stapler, like the one in Fig. 55, is excellent for nailing cabinet backs. This type drives a long staple, setting it

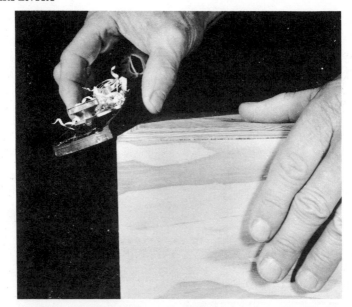

Fig. 53. A flush cabinet back can be beveled with a plane to make it less noticeable.

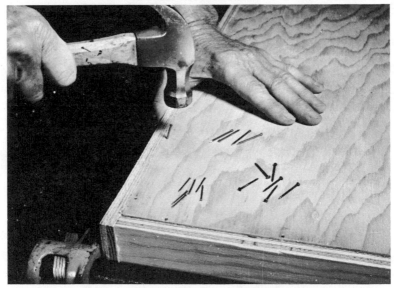

Fig. 54. Brads or blued lath nails can be used to fasten a back to the rabbeted cabinet sides.

below the surface if desired, and greatly speeds up the work. These staplers are sometimes available on loan or rental.

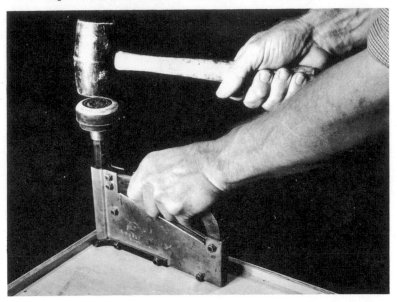

*Fig. 55. A heavy-duty stapler can be used to fasten the plywood back to cabinets.*

## SMALL BUILDING CONSTRUCTION

*(Information and illustrations courtesy Southern Pine Association)*

In constructing any small building, important principles to consider include rigid framing, insulation, stabilized floors, vibration prevention, and other features which contribute to proper construction.

No part of a building deserves more consideration than the selection of material for the framework. This is the hidden skeleton (Fig. 56) which must provide strength, stiffness, and permanence safeguarded by material that is properly seasoned.

Lumber as it comes from the tree is literally saturated with water. The process of removing the excess water is known as *seasoning*. It may be done by air drying over a long period of time, or by kiln drying. As moisture is removed, the wood shrinks. When "green" or unseasoned lumber is used, the shrinkage takes

169

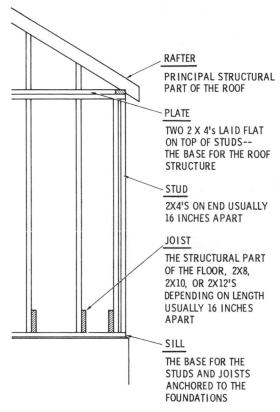

RAFTER
PRINCIPAL STRUCTURAL
PART OF THE ROOF

PLATE
TWO 2 X 4's LAID FLAT
ON TOP OF STUDS--
THE BASE FOR THE ROOF
STRUCTURE

STUD
2X4'S ON END USUALLY
16 INCHES APART

JOIST
THE STRUCTURAL PART
OF THE FLOOR, 2X8,
2X10, OR 2X12'S
DEPENDING ON LENGTH
USUALLY 16 INCHES
APART

SILL
THE BASE FOR THE
STUDS AND JOISTS
ANCHORED TO THE
FOUNDATIONS

*Fig. 56. The hidden skeleton of a frame building.*

place in the building and may cause cracks in the plaster and open joints. Shrinkage in the lumber should take place before the lumber is used and not after.

Great care is taken in producing high-quality lumber to see that the moisture in the wood is reduced to the proper point before the lumber reaches its final stage of manufacture. Scientific instruments enable the producer to determine when lumber is correctly seasoned.

### Good Construction Fundamentals

The weight of a small building is not adequate to hold it firmly on the foundation. Therefore, the *sills,* which rest on the foundation and support the walls and first floor joists, should be anchored to the foundation by bolts embedded in the concrete (Fig. 57). A 2″ × 6″ sill is recommended for one-story houses and two 2″ × 6″s for two stories. When placing the sills, mortar is first spread on the foundation and the sill tapped lightly to an even

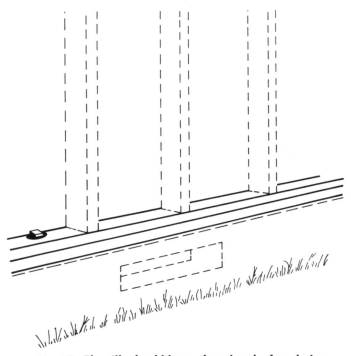

**Fig. 57. The sills should be anchored to the foundation.**

bearing. (Where there is no basement, the foundation walls should provide enough vents to insure free circulation of air throughout the area beneath the first floor.)

*Floor joists* are the most important part of the frame (Fig. 58). They support the floor and the entire contents of the building. Joists should be placed 16″ on center and possess enough stiffness not to be springy. Maximum strength and stiffness are obtained when the lumber is properly seasoned. Quality is particularly important in floor joists.

171

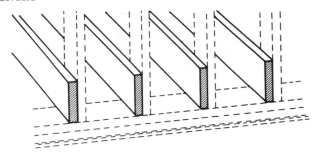

**Fig. 58. Floor joists support the floor and the contents of the building.**

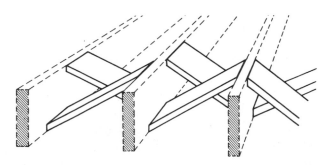

**Fig. 59. Cross-bridging is used to increase the rigidity of the floor and prevent vibration.**

*Cross bridging* (Fig. 59) is a must in good construction. The bridging consists of bracing with 1″ × 3″ or 1″ × 4″ pieces between the joists at intervals of 8′. These pieces should be double nailed at each end. Applied this way, cross bridging increases the rigidity of the floor and eliminates vibration. Joists which support partitions should be doubled (Fig. 60) and separated by solid wood bridging. Where the partition carries a load overhead, use three joists.

*Studs* (Fig. 61) support the ceiling and roof structure, and provide the base for the outside sheathing and the inside wall surfacing. The studs are placed on sills and are usually 2″ × 4″′s spaced 16″ on center. Studs should be doubled at the sides of all openings and tripled at each corner or angle of the structure. Joists placed next to studs should be securely nailed thereto.

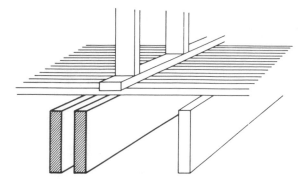

**Fig. 60. Joists should be doubled under partition walls.**

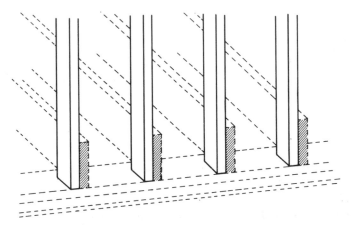

**Fig. 61. Studs support the ceiling and roof, and provide a base for outside sheathing and the inside wall surface.**

When studs are cut for windows or doors, the partition above the opening must be supported by a *header*. This may be two pieces of 2″ × 4″ placed on edge for small openings, or larger pieces or trusses for wider openings. *Sills* are placed at the bottom of the opening. See Fig. 62.

*Sheathing* should be applied to all outside walls. Sheathing insulates and strengthens the wall and provides the base for finish siding. Wood sheathing installed diagonally (Fig. 63) makes a very strong and stiff wall. Frequently, however, it is run hori-

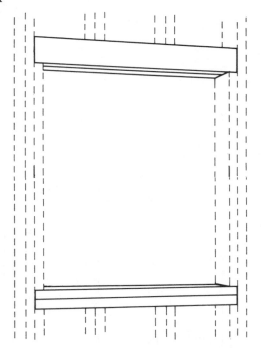

**Fig. 62. Headers are placed at the top of a window opening and sills at the bottom.**

zontally. Where this is done, the frame should be braced by a 1″ × 4″ piece let into the face of the stud at an angle of 45 degrees (Fig. 64) and extended from the sill to the rafter in one-story buildings; or to the second-floor joists in a two-story structure. With either method, the sheathing should extend to the foundation and be nailed to the joist ends and sills. Heavy waterproof felt should be applied over wood sheathing, tightly fitted around all openings, and securely nailed to the sheathing throughout the surface area with broad-headed roofing nails.

Insulating sheathing is being used in more and more modern construction. This is a fibered product heavily coated with asphalt on both sides. It is available in thicknesses of 1/2″ and 25/32″ and in 2″ × 8″ and 4″ × 8″ sheets. Waterproof felt is unnecessary with this type of sheathing. Its two major advantages are its insulating qualities and its speed of installation.

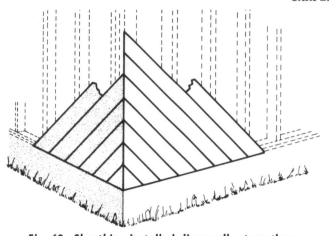

Fig. 63. *Sheathing installed diagonally strengthens the wall.*

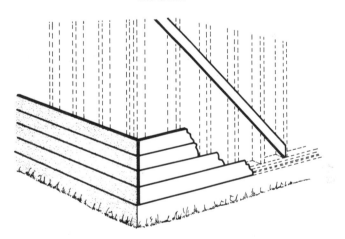

Fig. 64. *Where sheathing is applied horizontally, a 1" × 4" brace should be let into the studs at a 45° angle.*

The *subfloor* (Fig. 65) adds insulation and strength to the structure, and helps deaden sound. It is strongest when laid diagonally, which also permits the finish floor to be laid in either direction. The subfloor should extend between the studding to the outside wall sheathing and, on the first floor, be laid in an opposite direction to the way it is laid on the second floor. The

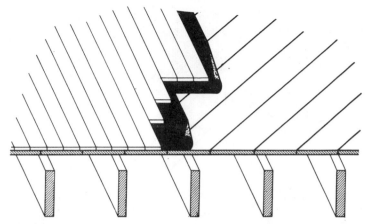

**Fig. 65. Subfloor is best laid diagonally and separated from the finish floor with building paper or asphalt felt.**

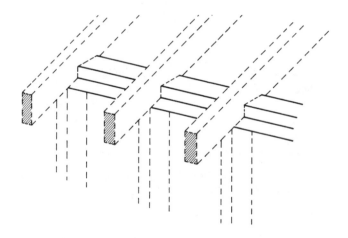

**Fig. 66. The top plate supports the roof rafters.**

material used should be dressed and matched lumber, preferably 1″ × 6″, securely nailed to each joist to prevent squeaking. Building paper or floor felt with a 4″ lap should be laid over the subfloor before proceedings with the finish floor.

The *plate* (Fig. 66) on top of the studs supports the roof and should be of the same width as the studs. The plate consists of

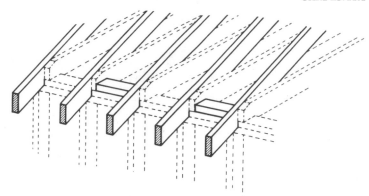

**Fig. 67. Anchor blocks used to tie the rafters to the plate.**

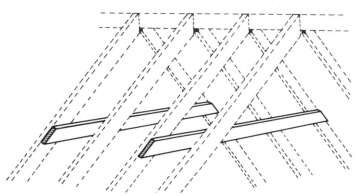

**Fig. 68. Collar beams are used to strengthen the roof.**

two thicknesses, one of which overlaps the end joints of the other. The member directly on top of the stud should be securely nailed to each stud.

Where *ceiling joists* are spaced the same as rafters, they should be nailed solidly to the rafters. The roof should be anchored to the side wall. An approved method, shown in Fig. 67, consists of anchor blocks of the same width as the plates, fitted alternately between the rafters and securely nailed to the rafters and plates.

Extra stiffness and strength for the roof are provided by placing *collar-beam* braces (Fig. 68) on every second pair of opposite rafters at about middle attic height, and securely spiked at each end. The material should be of the same dimensions as the rafters.

## Table 1. Your Guide to New Sizes of Lumber

| WHAT YOU ORDER | WHAT YOU GET | | WHAT YOU USED TO GET |
|---|---|---|---|
| | * Dry or Seasoned | ** Green or Unseasoned | Seasoned or Unseasoned |
| 1 x 4 | ¾ x 3½ | $^{25}/_{32}$ x 3$^{9}/_{16}$ | $^{25}/_{32}$ x 3⅝ |
| 1 x 6 | ¾ x 5½ | $^{25}/_{32}$ x 5⅝ | $^{25}/_{32}$ x 5½ |
| 1 x 8 | ¾ x 7¼ | $^{25}/_{32}$ x 7½ | $^{25}/_{32}$ x 7½ |
| 1 x 10 | ¾ x 9¼ | $^{25}/_{32}$ x 9½ | $^{25}/_{32}$ x 9½ |
| 1 x 12 | ¾ x 11¼ | $^{25}/_{32}$ x 11½ | $^{25}/_{32}$ x 11½ |
| 2 x 4 | 1½ x 3½ | 1$^{9}/_{16}$ x 3$^{9}/_{16}$ | 1⅝ x 3⅝ |
| 2 x 6 | 1½ x 5½ | 1$^{9}/_{16}$ x 5⅝ | 1⅝ x 5½ |
| 2 x 8 | 1½ x 7¼ | 1$^{9}/_{16}$ x 7½ | 1⅝ x 7½ |
| 2 x 10 | 1½ x 9¼ | 1$^{9}/_{16}$ x 9½ | 1⅝ x 9½ |
| 2 x 12 | 1½ x 11¼ | 1$^{9}/_{16}$ x 11½ | 1⅝ x 11½ |
| 4 x 4 | 3½ x 3½ | 3$^{9}/_{16}$ x 3$^{9}/_{16}$ | 3⅝ x 3⅝ |
| 4 x 6 | 3½ x 5½ | 3$^{9}/_{16}$ x 5⅝ | 3⅝ x 5½ |
| 4 x 8 | 3½ x 7¼ | 3$^{9}/_{16}$ x 7½ | 3⅝ x 7½ |
| 4 x 10 | 3½ x 9¼ | 3$^{9}/_{16}$ x 9½ | 3⅝ x 9½ |
| 4 x 12 | 3½ x 11¼ | 3$^{9}/_{16}$ x 11½ | 3⅝ x 11½ |

*19% Moisture Content or under.
**Over 19% Moisture Content.

## Table 2. Screw Sizes and Pilot-Hole Dimensions For Various Thicknesses of Plywood

| Plywood Thickness | Screw Size Dia. Length | Pilot-Hole Size |
|---|---|---|
| ¾" | #8-1½" | $^{5}/_{32}$" |
| ⅝" | #8-1¼" | $^{5}/_{12}$" |
| ½" | #6-1¼" | ⅛" |
| ⅜" | #6-1" | $^{7}/_{64}$" |
| ¼" | #4-¾" | |

## Table 3. Number of Board Feet in Various Lumber Sizes

| Size of piece (inches) | Length of piece (feet) | | | | | | | | |
|---|---|---|---|---|---|---|---|---|---|
| | 8 | 10 | 12 | 14 | 16 | 18 | 20 | 22 | 24 |
| 2 x 4 | 5⅓ | 6⅔ | 8 | 9⅓ | 10⅔ | 12 | 13⅓ | 14⅔ | 16 |
| 2 x 6 | 8 | 10 | 12 | 14 | 16 | 18 | 20 | 22 | 24 |
| 2 x 8 | 10⅔ | 13⅓ | 16 | 18⅔ | 21⅓ | 24 | 26⅔ | 29⅓ | 32 |
| 2 x 10 | 13⅓ | 16⅔ | 20 | 23⅓ | 26⅔ | 30 | 33⅓ | 36⅔ | 40 |
| 2 x 12 | 16 | 20 | 24 | 28 | 32 | 36 | 40 | 44 | 48 |
| 2 x 14 | 18⅔ | 23⅓ | 28 | 32⅔ | 37⅓ | 42 | 46⅔ | 51⅓ | 56 |
| 3 x 6 | 12 | 15 | 18 | 21 | 24 | 27 | 30 | 33 | 36 |
| 3 x 8 | 16 | 20 | 24 | 28 | 32 | 36 | 40 | 44 | 48 |
| 3 x 10 | 20 | 25 | 30 | 35 | 40 | 45 | 50 | 55 | 60 |
| 3 x 12 | 24 | 30 | 36 | 42 | 48 | 54 | 60 | 66 | 72 |
| 3 x 14 | 28 | 35 | 42 | 49 | 56 | 63 | 70 | 77 | 84 |
| 4 x 4 | 10⅔ | 13⅓ | 16 | 18⅔ | 21⅓ | 24 | 26⅔ | 29⅓ | 32 |
| 4 x 6 | 16 | 20 | 24 | 28 | 32 | 36 | 40 | 44 | 48 |
| 4 x 8 | 21⅓ | 26⅔ | 32 | 37⅓ | 42⅔ | 48 | 53⅓ | 58⅔ | 64 |
| 4 x 10 | 26⅔ | 33⅓ | 40 | 46⅔ | 53⅓ | 60 | 66⅔ | 73⅓ | 80 |
| 4 x 12 | 32 | 40 | 48 | 56 | 64 | 72 | 80 | 88 | 96 |
| 4 x 14 | 37⅓ | 46⅔ | 56 | 65⅓ | 74⅔ | 84 | 93⅓ | 102⅔ | 112 |
| 6 x 6 | 24 | 30 | 36 | 42 | 48 | 54 | 60 | 66 | 72 |
| 6 x 8 | 32 | 40 | 48 | 56 | 64 | 72 | 80 | 88 | 96 |
| 6 x 10 | 40 | 50 | 60 | 70 | 80 | 90 | 100 | 110 | 120 |
| 6 x 12 | 48 | 60 | 72 | 84 | 96 | 108 | 120 | 132 | 144 |
| 6 x 14 | 56 | 70 | 84 | 98 | 112 | 126 | 140 | 154 | 168 |
| 8 x 8 | 42⅔ | 53⅓ | 64 | 74⅔ | 85⅓ | 96 | 106⅔ | 117⅓ | 128 |
| 8 x 10 | 53⅓ | 66⅔ | 80 | 93⅓ | 106⅔ | 120 | 133⅓ | 146⅔ | 160 |
| 8 x 12 | 64 | 80 | 96 | 112 | 128 | 144 | 160 | 176 | 192 |
| 8 x 14 | 74⅔ | 93⅓ | 112 | 130⅔ | 149⅓ | 168 | 186⅔ | 205⅓ | 224 |
| 10 x 10 | 66⅔ | 83⅓ | 100 | 116⅔ | 133⅓ | 150 | 166⅔ | 183⅓ | 200 |
| 10 x 12 | 80 | 100 | 120 | 140 | 160 | 180 | 200 | 220 | 240 |
| 10 x 14 | 93⅓ | 116⅔ | 140 | 163⅓ | 186⅔ | 210 | 233⅓ | 256⅔ | 280 |
| 10 x 16 | 106⅔ | 133⅓ | 160 | 186⅔ | 213⅓ | 240 | 266⅔ | 296⅓ | 320 |
| 12 x 12 | 96 | 120 | 144 | 168 | 192 | 216 | 240 | 264 | 288 |
| 12 x 14 | 112 | 140 | 168 | 196 | 224 | 252 | 280 | 308 | 336 |
| 12 x 16 | 128 | 160 | 192 | 224 | 256 | 288 | 320 | 352 | 384 |
| 14 x 14 | 130⅔ | 163⅓ | 196 | 228⅔ | 261⅓ | 294 | 326⅔ | 359⅓ | 392 |
| 14 x 16 | 149⅓ | 186⅔ | 224 | 261⅓ | 298⅔ | 336 | 373⅓ | 410⅔ | 448 |
| 16 x 16 | 170⅔ | 213⅓ | 256 | 298⅔ | 341⅓ | 384 | 426⅔ | 469⅓ | 512 |

## Table 4. Types of Glue Suitable for Plywood

| TYPE OF GLUE | DESCRIPTION | RECOMMENDED USE | PRECAUTIONARY USE |
|---|---|---|---|
| HIDE GLUE | Comes as flakes to be heated in water, or in prepared form as liquid hide glue. Very strong, tough, light color. | Excellent for furniture and cabinetwork. Gives strength even to joints that do not fit very well. | Not waterproof; do not use for outdoor furniture or anything exposed to weather or dampness. Clamp 3 hours. |
| UREA RESIN GLUE | Comes as powder to be mixed with water and used within 4 hours. Light colored. Very strong if joint fits well. | Good for general wood gluing. First choice for work that must stand some exposure to dampness, since it is almost waterproof. | Needs well-fitted joints, tight clamping, and room temperature 70° or warmer. Allow 16 hours drying time. |
| LIQUID RESIN (WHITE) GLUE | Comes ready to use at any temperature. Clean-working, quick-setting. Strong enough for most work, though not quite so tough as hide glue. | Good for indoor furniture and cabinetwork. First choice for small jobs where tight clamping or good fit may be difficult. | Not sufficiently resistant to moisture for outdoor furniture or outdoor storage units. Sets in 1½ hours. |
| RESORCINOL (WATERPROOF) GLUE | Comes as powder plus liquid, must be mixed each time used. Dark colored, very strong, completely waterproof. | This is the glue to use with exterior type plywood for work to be exposed to extreme dampness. | Expense, trouble to mix and dark color make it unsuited to jobs where waterproof glue is not required. Allow 16 hours drying time. |

CHAPTER 5

# Roofing Maintenance

Building owners and maintenance personnel have a responsibility to keep the roof in good repair so as to prevent possible excessive expenditure to building sections, such as sheathing, and to the building contents by roof leakage. The roof should be inspected at least once a year and any necessary repairs made.

## SKYLIGHTS

Skylights are always a maintenance problem. A leak which occurs in a roof skylight is often difficult to trace. Leaks occur when glazing compound (putty) becomes deteriorated and crumbles, leaving points for condensation or rain to enter. It is best to remove the putty entirely if it is old and crumbled, then clean the window frame, apply linseed oil with a brush (particularly in wooden sash) and reinstall the glass. The new rubber calking compounds work exceptionally well and outlast ordinary putty by years.

If sections of the skylight are rotted (or, if metal, are rusted), these damaged sections must be repaired with putty, plastic wood, metal or plastic roofing cement, or rubber-base calking cement.

The skylight framing should be inspected and the required repairs made. Often debris, such as leaves, will collect and will cause water to stand around the skylight, resulting in additional leak hazards. All debris must be periodically removed. Exposed nails or poor seams must be checked and roofing cement applied to them.

Flashing failures often occur at skylights, parapet walls, and vents, and these points should be inspected carefully. In most cases, the application of plastic roofing cement will insure a satisfactory, though temporary, repair.

To repair felt-base flashing, force plastic cement behind the felt and then seal the edge with a 5-inch strip of glass-type or bituminous-saturated cotton fabric embedded into the cement. Finish with a coat of plastic or regular roofing cement.

Occasionally, it is necessary to place a new felt flashing over the entire base flashing. Place the new flashing into a bed of previously applied asphalt plastic cement or cold-application roofing; set it firmly and then coat the new covering with plastic or cold-application roofing again. Hot asphalt is often used for this type of repair.

When skylight repairs are being analyzed, consideration should be given to the possibility of complete removal of the skylight if it no longer serves a useful purpose.

## METAL FLASHINGS

Metal flashing is used with cement-asbestos, slate, tile, metal, and wooden-shingle roofs, as well as with mineral-surfaced and built-up roofing. Expansion and contraction in the metal often tends to break the joints open and loosen the fasteners. Commonly used metals for roof flashings are galvanized iron, copper, and aluminum.

### Copper Flashing

Copper flashing and counter flashing is more expensive than other types, but requires little repair. Broken seams must be re-soldered and loose sections refastened.

### Aluminum and Galvanized-Iron Flashings

Aluminum flashing must be inspected for loose sections, which should be refastened and re-seamed if necessary. Galvanized-iron flashing has a tendency to rust after continued exposure. A rust-inhibitive paint or roofing asphalt should be applied to prevent continued rusting.

In all areas where a roof angle meets another, or where metallic sheets have been applied, such as at plumbing vents, chimneys, and parapet walls, leaks occasionally develop. Inspect these areas, and if the flashing is not loose, apply a generous coating of asphalt cement. If the flashing has become loose and protrudes away from the brickwork, form a ball or wedge of lead wool and drive it into the brick joints where the flashing is loose. Then apply asphalt cement.

In making repairs to valleys, care must be taken so as not to interrupt the flow of water. The valley is the flow point for the roof drainage system and during very heavy rains, interruption of the water flow could result in the water backing up under the shingled roof areas. Fig. 1 shows typical flashing details where a roof meets a chimney.

## SLATE ROOFS

The most common repair necessary on slate roofs is the replacement of a damaged or broken slate. In order to remove the cracked slate, cut the nails holding it in place with a hack saw blade or a roofing tool known as a ripper. Place a new slate into

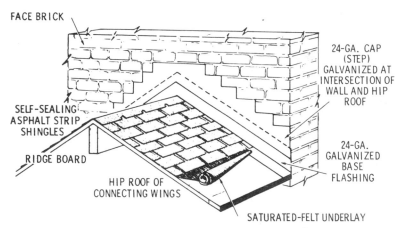

*Fig. 1. Typical chimney flashing detail.*

the location where the old one has been removed, and nail it through the vertical joint of the course above it. Place a 4″ × 6″ piece of aluminum or copper strip under the course above the nail. This will prevent leaks at the nail hole. It is also advisable to place a very small amount of plastic asphalt cement over the nail head (Fig. 2).

Occasionally a leak may occur when high winds blow snow or rain under the lower edge of the slate. One method of repairing this type of leak is to raise the bottom edge of each slate just enough to place some asphalt plastic roofing cement underneath. Press the slate into place. This method seals each slate course to the one underneath it.

## TILE AND CEMENT-ASBESTOS ROOFS

Broken shingle tiles may be replaced in a manner similar to that used for slate roofs. On Spanish tile, trowel a mixture of Port-

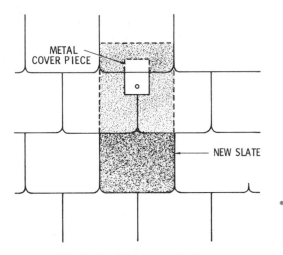

*Fig. 2. Replacing a new slate.*

land cement mortar on the surface of the tile being placed, and also on the location at which the tile will rest on the course below. Press the new tile firmly into place and wipe away any excess mortar.

Cement-asbestos shingles are normally applied in one of three different ways—the American, hexagonal, and Dutch-lap methods. For the American method, the damaged shingles can be replaced

184

in the manner as for slate. For the other two methods, cut the metal fasteners with a hack-saw blade or ripper, remove the damaged shingle, and place new fasteners and shingles. Apply plastic roofing cement if required. The newly developed rubber-type seal used for calking will blend better on cement-asbestos shingles if this type patching is necessary. Rubber-base cement will last longer than the conventional roofing cement. Figs. 3 and 4 show the installation procedures for hexagonal-type shingles.

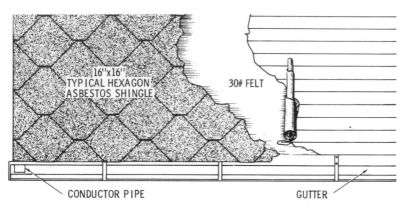

**Fig. 3. Installation details of 16" × 16" hexagon cement-asbestos shingles.**

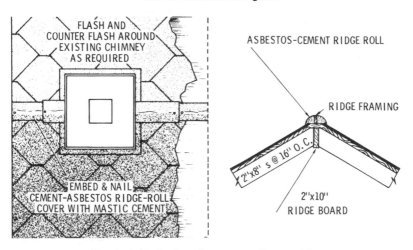

**Fig. 4. Installation of cement-asbestos ridge.**

185

## COPPER ROOFS

Copper roof leaks generally occur due to contraction and expansion of the metal, causing the seams to break. These broken seams can be repaired by soldering. If leaks occur from small holes, these also can be soldered. Where soldering is necessary, scrape the surfaces to a bright finish by using a sharp scraper or emery cloth. Wipe the surfaces clean and apply soldering flux, and then solder with 50-50 (50% lead and 50% tin) solder. Using too much solder or an iron that is not hot enough will create an unsightly job.

If copper decking has been installed without provision for contraction and expansion, joints must be installed or contraction and expansion breaks will continue to occur.

## GALVANIZED-IRON ROOFS

Galvanized-iron roofs are subject to rust and must be painted periodically in order to prevent damage. Seams also separate and leak. If necessary to paint this type of roof, and it has not weathered for at least 6 months, the surface should be cleaned with a solution of 25% vinegar and 75% water to remove the greasy film. Paint can be applied after loose nails have been driven into place and the roof surface cleaned. In some sections of the United States, coating galvanized metal roofs with liquid asphalt roof coating of the type normally sold by paint stores, hardware dealers, and lumber yards, has been very successful.

## CONCRETE ROOFS

Leaks often develop in concrete roofs primarily because of faulty cement mixes which leave an overly-porous surface that absorbs moisture. Cracks may also develop from expansion and contraction, unusual vibration, and settling.

Repairs are generally made with asphaltic materials. If the repair requirements are negligible, the entire surface can be coated with a cold application of liquid asphaltic roofing material. If the cracks and repairs are extensive, it is best to consult a reliable roofing contractor.

## BUILT-UP ROOFS

Built-up roofs consist of layers of tarred felt cemented with hot tar or asphaltic pitch to form a continuous skin or layer over the entire roof deck. The number of layers, or plies, of tarred felt varies from 3 to 5, and depends on the type of roof decking and length of desired service. The number of plies does not affect its water resistance, but the greater number of plies does contribute to longer life.

On the average, 5-ply roofs are guaranteed for 20 years, 4-ply for 15 years, and 3-ply for 10 years. This average guarantee does not always hold true, as some roofing contractors will guarantee 4-ply gravel or slag roofs for 20 years and 3-ply for 15 years, depending on the type and incline of the roof.

### Blisters or Bulges

Built-up roofing should be inspected at least once a year, preferably in the late summer or early fall. If blisters or bulges are found, these can be repaired as follows:

1. Cut the blister open, using an X-cut. Fold back the flaps and allow to dry thoroughly.
2. Trowel a good plastic roofing cement over the exposed area. Replace the flaps and press into place.
3. Brush over the area and the adjacent perimeter with cold-application roof coating, or with plastic roofing cement if the area is badly weathered. Place bitumin-impregnated cotton fabric over the area.
4. After the fabric settles or is pushed into place, brush again with liquid asphalt coating and/or a hot-application type of roofing material. Most roof repairs are made with a cold-application type of roof coating of the consistency as it comes from the can.

### Cracks, Tears, and Open Seams

Trowel asphaltic cement over the crack or seam after cleaning and nailing (if required). Then imbed bitumin-impregnated fabric and brush thoroughly with liquid asphalt and/or a hot-application type of roof coating. For small areas, a cold-application type

187

of roof coating, such as obtained from hardware stores or lumber yards, is recommended.

### Recoating

Over a period of years, oils in felts have a tendency to dry out, leaving the felt in a porous, brittle condition. Cracks or breaks may result. Due to these conditions, all roofs of this type should be recoated with asphaltic liquid roof coating or hot tar at least every five years, or as conditions warrant.

Prior to the application of the asphaltic coating, all cracks and breaks should be repaired with asphaltic plastic cement. Set all loose nails and remove debris from the roof. Brush the roofing material over the felt surface. On large surfaces, spray equipment may be used if handled by a reliable roofing contractor or other personnel skilled in the use of this type of equipment.

Care should be used in the selection of roof coating materials, as different grades of asphalt do exist and some inferior materials are sold. Secure your materials from a reliable dealer who also can give advice on roofing problems.

### Gravel-Surface Roof Repair

Built-up gravel-surface roofs can be repaired in the same manner as smooth-surface roofs, excepting that loose gravel or slag must be swept off the area being repaired. After the liquid roofing material is applied and allowed to settle, the gravel or slag can be redistributed to its original location.

## WOODEN SHINGLES

Badly split or deteriorated wooden shingles should be replaced. First split the shingle to be removed into several pieces with a chisel, being careful not to damage other surrounding shingles. If the nails above the broken shingle must be removed, this can be done by sawing with a hacksaw blade, cutting with a shingle hook, or removing with a nail puller. After the preparation is complete, drive the new shingle into place and fasten it to the roof by nailing through the new shingle between the two shingles immediately above it. Place a thin coating of roofing cement over the nail and then drive a strip of copper or aluminum metal over the area so that the metal rests over the exposed nailhead, similar to the procedure used for repairing slate roofs.

If an entire reroofing is being considered, then all of the existing wooden shingles should be checked for firmness, nailing down those that are loose. Wooden shingles or cedar "shakes" can be placed directly over the old wooden shingles, saving labor and providing additional insulation for the building.

Wood shingles may be laid on slopes as gentle as 1 in 4 but perform best on steeper slopes.

The life of wooden shingles or shakes can be extended by applying paints or stains which contain creosote and other preservatives. For long roof life a preservative paint or stain should be applied about every five years.

Wood shingles are usually laid over a solid deck particularly in colder climates. However, in warm, moist climates, wood shingles laid over slats allow air to flow from the attic space thus reducing the high temperature and excessive moisture problems.

For new construction, Table 1 on page 195 will be found useful to determine the approximate number of nails and shingles that will be required to apply wood shingles.

## Felt Roll Roofing

If the entire roof appears weatherbeaten, with only a few small holes, patch the holes with plastic roofing cement and nail down any loose roofing material. After repairs have been made, brush over the entire roof with a cold application of asphaltic roofing material. It is good practice to inspect roofs of this type once a year and, if necessary, patch with roofing cement as required. It is recommended that the entire roof be coated with liquid roof coating at least once every four years, or as conditions may warrant.

## GUTTER REPAIRS

Low spots often develop in gutters, resulting in water pooling at the low points. The pitch or grade of the gutter can often be corrected (reset), so the water will flow toward the downspout or conductor pipe, by adjusting the gutter brackets.

Gutter leaks can be repaired by placing a very thin application of roofing asphalt over the hole, and then firmly pressing a fairly heavy piece of aluminum foil (TV snack-tray pieces work well) or a piece of other light metal or felt or asphalt-saturated cotton fabric into place. Coat over the material with more asphalt cement, being careful not to build a dam in the gutter. If the gutter is rusting, then the entire gutter should be cleaned, brushed, and painted. If the gutter is badly rusted, an application of liquid roofing asphalt works well in place of paint. It is recommended that copper gutters be soldered at all leak points.

Wooden gutters can be repaired by plugging small holes with a wooden plug and then coating with plastic wood or a rubber-base calking compound. If there is a split in the wood, it can be repaired with calking compound, preferably one with a rubber base.

## MINERAL-SURFACED ASPHALT SHINGLES

### Repair

For a quick repair of cracked or damaged asphalt shingles, apply a liberal coating of plastic roofing cement. If possible, lift the shingle up and apply the bituminous plastic both under and into the cracked or damaged opening. Often small cracks and holes appear in roofing from stones which are thrown onto the roof.

If it is necessary to remove an asphalt shingle, cut it away, remove the protruding nails, and place a new shingle into place. Cover the old holes and nail heads with plastic cement.

### Application

New mineral-surfaced asphalt shingles can be laid over an old roof of wooden or asphalt shingles with good results. Care must be used in the selection of nail lengths so that the nails penetrate into, but not through, the roof sheathing. It is not recommended to apply asphalt shingles to roofs which have a pitch of less than 3″ per horizontal foot. (See Fig. 5.)

If old shingles (either wood or asphalt) are loose or curled, care should be taken to nail them securely in place. If the shingles or the wooden deck are not sound, the shingles should be removed and the deck repaired. It is suggested that beveled wood strips be nailed just below the butts of the old wooden shingles to improve the appearance of the finished roof.

In order to prevent condensation of moisture under the shingles, place 15- or 30-pound asphalt-saturated felt over the wood deck before placing the shingles. The felt should be laid horizontally, with at least a 2″ lap, with sufficient nails to hold it in place. Metal edging strips (Fig. 6) are recommended on new roof construction to improve the appearance.

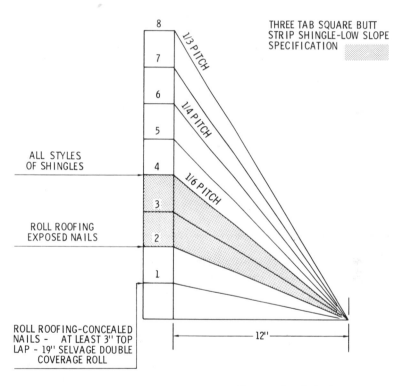

*Courtesy Georgia-Pacific Corporation.*

**Fig. 5. Minimum pitch requirements for asphalt roofs.**

## Valleys

To ensure good protection when reroofing or when a new roof is applied, two strips of 90-pound mineral-surfaced roofing material should be used at all valley points. The first strip, cut 18″ in width, is laid with the mineral surfacing downward, and the second strip, cut 36″ wide, with the mineral surfacing turned up. After the material has conformed to the roof, nail the first strip at approximately 18″ intervals, and at the same approximate intervals for the second strip, staggering the nails so as not to have the same locations as the first nailing. Chalk lines for the placing of roof shingles should be at least 3″ from the center of the valley material at the ridge and 4″ from the center of the valley at the eave. Keep all nails at least three inches from the chalk lines when applying the shingles.

### Starters

It is general practice to apply inverted shingles along the roof eave, with the ends butting and the long side horizontal. Often 90-lb. mineral-surfaced roll roofing cut to a width of 18″ is used for a starting strip. The lower end of the strip should extend 1/8″ out over a metal edging strip. Do not extend the starting strip into the valley strip. Nail the starter strip about 2-1/2″ above the lower edge of the eave, placing the nails so they will not be exposed when the shingles are laid.

The first few courses of shingles should be applied by standing on staging supported by ladders placed against the eaves or edge of the building at a safe angle of 70 degrees. The legs of the ladders should be set firmly so they cannot slip. In order to secure the proper horizontal alignment of the shingles, chalk lines should be struck at measured distances from the eave as the shingles are laid. Most shingles are self aligning. It is recommended that the manufacturer's instructions be followed.

The proper placement of the nails is generally indicated in the manufacturer's instructions for the particular type of mineral-surfaced asphalt shingles being installed. Roofing nails are generally 10-1/2 gauge, but should not be thinner than 12 gauge, with heads not less than 3/8″ in diameter. For standard-weight shingles on new work, a 1″ nail is used; for large or heavy shingles, 1-1/4″ nails, and for reroofing over wood shingles, 1-3/4″ nails. The nails should not be driven into knotholes or cracks, and should not be countersunk.

## American Method Shingles

The American method of shingle arrangement is in general use. This type of shingle combines 3 or 4 shingles into one piece for easy application, reducing labor costs, insuring a tighter roof, and to ensure better self spacing and alignment of the shingles. This type of shingle is generally referred to as the *three-in-one* or *four-in-one* mineral-surfaced shingle. Complete directions for installing are normally found in the shingle bundles or packages.

## Wind Damage

In high wind or excessive storm areas, it is best to use thick-butt or self-adhesion type of mineral-surfaced shingles. In the self-adhesion type of shingle, the lower edge of each shingle is precoated with asphalt and sticks to the shingle beneath. They should be pressed down during application. Adhesion is greater if the shingles are applied during warm weather when the plastic material at the end of the tabs is soft and the weight of the roofing material assures the sticking of one shingle to another. The self-lock type of roofing shingle is also used with success in high-wind areas.

## ROLL ROOFING

It is recommended that the pitch of the roof should not be less than 3″ per horizontal foot when applying smooth or mineral-surfaced roll roofing. The roof should be dry and all knotholes covered with tin or aluminum strips. Roll roofing can be applied over old wood shingles if they are not badly curled, and both the deck and shingles are sound. Prior to applying roll roofing over old shingles, nail down all curled shingles and replace those that are missing. To improve the appearance of the finished roof, use beveled wood strips nailed just below the butts of the old shingles, and of the same thickness as that of the butt of the shingles.

To ensure a better finished job when applying materials to a new roof, metal edging strips are recommended, as shown in Fig. 6. Allow the strips to project about 1/2″ beyond the edge of the roof, and nail in place.

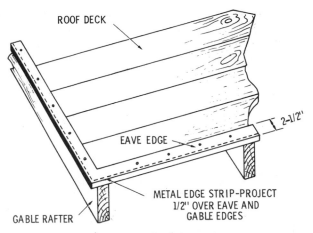

**Fig. 6. Metal edging strips.**

Double thicknesses of roofing should be applied in all valleys. The first strip should be 18″ wide with the mineral surfacing down, and the second strip 36″ wide with the mineral surfacing up. After it has been determined that the strips are laying smooth and in contour with the roof surface, nail each strip at 18″ intervals. As a guide for cutting roofing paper, strike chalk lines 12″ from the center of the valley from the eave to the ridge.

## Laying and Nailing

In order to accomplish a desired roofing job, roofing should not be installed at temperatures below 40 degrees F. If installed in cool weather, store the materials in a warm place, and when applying, unroll the material carefully to prevent cracking and breaks. Allow the material to flatten before cutting.

Start at the eave or lower edge of the roof and work up, allowing the first sheet to project about 1/4″ over the edging strip. As the sheets are placed, allow an overlap of 2″, nailing the top at approximately 30″ centers and 1″ from the top section. Allow each sheet to lay smooth, applying a continuous coating of roofing cement to a 2″ lap of each sheet, and allow the cement to become slightly firm before finish nailing. All end laps and valleys should have a 6″ continuous coating of roofing cement.

On new work, use the nails furnished with the roofing, and on wood shingles, use regular roofing nails at least 1-1/2" long. Drive the nails straight and flush, 1" from the edge of the roofing and about 2-1/2" apart.

When placing roll roofing, face the laps away from the direction of the prevailing wind and cement all seams.

When applying flashing at brick work roof projections, rake the brick joint approximately 1" deep and place the top edge of the flashing into the opening between the brick. Metal counter flashing is recommended, although roofing material is sometimes used. After the flashing has been installed and fastened, apply a liberal coating of plastic roofing cement.

To fasten metal flashing into stone or brick "rake," some mechanics ball together lead wool of the proper size and pound it into place before applying plastic roofing cement.

### Ridges and Hips

General practice for ridges and hips is to apply a double thickness of material at least 18" wide. The first strip is scatter-nailed and the second nailed on 2-1/2" centers about 1" from the edges. Ridge and hip strips should be bedded into 2" widths of roofing or plastic cement before finish nailing. Before final nailing, allow the cement to become tacky or firm. Care should be used so as not to mar or break the material during application, particularly during hot weather.

### Table 1. Wood Shingle Roofing Requirements

| Laid to Weather | MATERIAL Per 100 Square Feet of Surface | | | NAILS per 100 Square Feet | |
| --- | --- | --- | --- | --- | --- |
| | Shingles per 100 Sq. Feet | Waste | Shingles per 100 Sq. Ft. with Waste | 3d Nails | 4d Nails |
| 4" | 900 | 10% | 990 | 3-¾ Pounds | 6-½ Pounds |
| 5" | 720 | 10% | 792 | 3 Pounds | 5-¼ Pounds |
| 6" | 600 | 10% | 660 | 2-½ Pounds | 4-¼ Pounds |

NOTE: Nails based on using 2 nails per shingle.

*Courtesy Georgia-Pacific Corp.*

195

## ROOFING TABLES

Roofing products are many and varied. To aid in the selection of the proper roofing materials, Table 2 is included. Table 3 lists the nail requirements for different types of asphalt roofing products.

Table 4 will be found useful to determine the area of the roof when only the plan of the building and pitch of the roof is given.

**Table 2. Asphalt Roofing Selection Data**

| PRODUCT | Approximate Shipping Weight per Square | Packages Per Square | Length | Width | Units Per Square |
|---|---|---|---|---|---|
| Giant Individual Dutch Lap | 162 lb. | 2 | 16" | 12" | 113 |
| Giant Individual American | 325 lb. | 4 | 16" | 12" | 226 |
| Individual Staple Down | 135 lb. | 2 | 16" | 16" | 80 |
| Individual Lock Down | 135 lb. | 2 | 16" | 16" | 80 |
| 2 and 3 Tab Hex Strip | 167 lb. | 2 | 36" | 11⅓" | 86 |
| 3 Tab Square Butt Strip Shingle | 210 lb. 262 lb. | 2 or 3 | 36" 36" | 12" 12" | 80 100 |
| 19" Selvage Double Coverage | 140 to 144 lb. | 2 | 36" | 36" | |
| Pattern Edge Roll | 105 lb. 105 lb. | 1 1 | 42' 48' | 36" 32" | |
| Mineral Surfaced Roll | 90 lb. 90 lb. 90 lb. | 1.0 | | | 1.0 1.075 1.15 |
| Smooth Roll | 65 lb. 55 lb. 45 lb. | 1 1 1 | 36" 36" 36" | 36" 36" 36" | |
| Saturated Felt | 15 lb. 30 lb. | ½ ¼ | 144' 72' | 36" 36" | |

Courtesy Georgia-Pacific Corp.

## Table 3. Nail Requirements For Asphalt Roofing Products

| Type of Roofing | Shingles per Sq. | Nails per Shingle | Length of Nail* | Nails per Square | Pounds per Square (approximate) 12 ga. by 7/16" head | 11 ga. by 7/16" head |
|---|---|---|---|---|---|---|
| Roll Roofing on new deck | | | 1" | 252** | .73 | 1.12 |
| Roll Roofing over old roof'g | | | 1¾" | 252** | 1.13 | 1.78 |
| 19" Selvage over old shing. | | | 1¾" | 181 | .83 | 1.07 |
| 3 Tab Sq. Butt on new deck | 80 | 4 | 1¼" | 336 | 1.22 | 1.44 |
| 3 Tab Sq. Butt reroofing | 80 | 4 | 1¾" | 504 | 2.38 | 3.01 |
| Hex Strip on new deck | 86 | 4 | 1¼" | 361 | 1.28 | 1.68 |
| Hex Strip reroofing | 86 | 4 | 1¾" | 361 | 1.65 | 2.03 |
| Giant Amer. | 226 | 2 | 1¼" | 479 | 1.79 | 2.27 |
| Giant Dutch Lap | 113 | 2 | 1¼" | 236 | 1.07 | 1.39 |
| Individ. Hex | 82 | 2 | 1¾" | 172 | .79 | 1.03 |

(*) Length of nail should always be sufficient to penetrate at least ¾" inch sound wood. Nails should show little, if any, below underside of deck.
(**) This is the number of nails required when spaced 2" apart.

## Table 4. Obtaining Roof Area From Plan Area

| Rise | Factor | Rise | Factor |
|---|---|---|---|
| 3" | 1.031 | 8" | 1.202 |
| 3½" | 1.042 | 8½" | 1.225 |
| 4" | 1.054 | 9" | 1.250 |
| 4½" | 1.068 | 9½" | 1.275 |
| 5" | 1.083 | 10" | 1.302 |
| 5½" | 1.100 | 10½" | 1.329 |
| 6" | 1.118 | 11" | 1.357 |
| 6½" | 1.137 | 11½" | 1.385 |
| 7" | 1.158 | 12" | 1.414 |
| 7½" | 1.179 | | |

The horizontal or plan area (including overhangs) should be multiplied by the factor shown in the column opposite the rise, which is given in inches per horizontal foot. The result will be the roof area.

197

# CHAPTER 6

# Glazing and Calking

At first thought, it might seem that glazing and calking would require the same knowledge and skill for both. There is a general similarity in the two job categories, but each has its own problems as far as maintenance is concerned.

## GLAZING

The word "glazing" would infer that glass maintenance is concerned only with replacing broken glass. This, however, is not true, although one of the duties of maintenance men is to replace broken glass in metal and wooden sash, doors, and partitions. Glazing maintenance also involves the replacement of loose, deteriorated, and missing putty or compound, as well as reconditioning window sash. It is advisable to replace the putty or compound with the same type of material that was originally used. However, many new glazing materials are now available, and a particular job may warrant making the repair with one of these new, more expensive, but longer-life compounds.

### Purchasing Glass

Glass for windows and doors is packaged in many ways, but the most economical way to purchase it is in boxes containing 50 square feet or more. Individual pieces can be obtained from most hardware or paint stores, where the dealer will cut the glass to any required size.

Bulk glass is available in widths from 6″ to 12″ in 1-inch intervals and from 12″ to 30″ in 2-inch intervals. Stock lengths are from 8″ to 40″ in 2-inch intervals. Glass dealers can supply other reasonable sizes on special order.

Specific requirements determine the type of glass to use, but the following schedule can be used for installing clear glass:

Use single-strength glass for areas up to 350 sq. in., double-strength glass for areas from 350 sq. in. to 6 sq. ft., and 3/16″ glass for areas over 6 sq. ft. where plate glass is not a requirement. In locations where glass is touched, pushed, or easily struck, double-strength is recommended for even the smaller areas.

## Shatterproof Glass

*Shatterproof glass* is recommended for locations such as engine test rooms, high-vibration areas, locations where injury may result from sharp fragments of glass, and in high-traffic locations where pedestrian traffic is constant. 1/4″ laminated glass, polished on both sides, is generally recommended.

## Polished Plate Glass

1/4″ glazing-quality, polished, clear-vision, *plate glass* is generally used in locations such as entranceways, large openings, and for large-size glass shelves. Plate glass used for shelving and counter tops should be beveled slightly and ground smooth. Glass used for small shelves and trays can be double-strength window glass, but all edges should be ground smooth.

## Wire Glass

*Wire glass* can be clear or figured, as well as flat or corrugated. In locations where security and clear vision are required, 1/4″ clear, flat, wire glass polished on both sides is used. In locations such as shops and garages where clear vision is not always required, 1/4″ figured or ribbed glass that is smooth on one side is used.

## Obscure Glass

In locations such as toilets, dressing rooms, and shower rooms, where clear vision is not required, *obscure glass* is used. Obscure glass is normally of good glazing quality, 1/4" thick, with a rolled figure on one side and smooth on the opposite side.

Many other types of glass, including *light-diffusing, heat-absorbing, colored,* and *lead glass* are available. Information concerning these types can be secured from a glass dealer. Installation procedures for almost all glass is nearly the same.

### Window Pane Installation

The first thing to be done when replacing a window pane is to remove the old or broken glass from the window sash. When working with glass, it is recommended that a pair of gloves be used. All of the old glass must be removed; if some pieces do not come loose with gentle tugging, remove the old putty holding it in place. When removing the old putty, use caution so that the wood is not gouged. Putty which is very hard and firm can be softened with the use of a hot soldering iron. Glazier's points which remain in the wood after glass removal should be taken out with a pair of pliers.

Almost all hardware and paint dealers will cut glass to the required size, but care should be used in measuring for the replacement. Measure the opening where the glass will be placed, and subtract 1/8" from the height and width to determine the proper replacement size. As an example, if the opening into which the glass is to be placed measures 12" × 18", the replacement glass size will be 11-7/8" × 17-7/8". The smaller size is necessary to allow for contraction and expansion of the wood with temperature and humidity changes.

When all old putty and glazier's points have been removed from the sash, clean the area carefully with a cloth or brush and apply a coat of linseed oil. The linseed oil will preserve the wood and prevent it from absorbing the oil from the putty, causing it to dry out too quickly.

Place a bed of putty 1/16" thick on the lip of all four sides of the sash groove. Press the glass into position and insert the glazier's points at intervals of 4" to 8" apart. Form a pencil-like roll of putty approximately 1/4" in diameter and several inches long. Place this strip of putty, starting at one corner of the sash and going completely around the new glass. Complete the job by forcing the putty into place with a putty knife held at a 45-degree angle. For most puttying jobs, the new type strip putty, which comes in rolls and is about an 1/8″ in diameter, works very well. Many of the tube-type rubber-base compounds also work very well, but are more expensive and are not generally used for setting glass in wooden sash. The putty should be allowed to set for several days before painting, but should be painted within a two-month period as it may dry out so much as to curl away from the glass and the wood. Many workmen prefer to use glazing compound instead of putty. This material does not dry as hard as putty and is therefore easier to remove when reglazing is necessary. Knife-grade compound should be used.

## Glass Replacement in Metal Sash

Replacing a piece of glass in an aluminum sash is more difficult than in a wooden sash. However, the usual way is to carefully remove the damaged or broken glass. Measure for the new glass from the bottom of the groove on one side to the bottom of the groove on the opposite side, subtracting 1/16″. Do this also for the other two sides of the sash, again deducting 1/16″. The 1/16″ allows for the thickness of the rubber window gasket.

Now remove one side and the top of the sash. In most aluminum sash, angle irons are located inside each mitered corner to hold the sash frame together. Pry the corners apart, using care not to damage the sash. Place the rubber gasket around the new piece of glass, fit it into position, place it into the sash grooves, and lock the sash together.

Vinyl and neoprene gasket channels are being used by some manufacturers, and a procedure similar to that used for the rubber gasket must be used. Wire clips are also used for glazing metal sash.

In other instances, a rubber-base compound, premixed and purchased by the tube, is being used for installing glass in aluminum and steel windows. Whatever procedure is used, a positive seal is required between the glass and metal on both sides of the glass.

1. Purchase and/or cut glass to the correct size.
2. Remove the broken glass.

3. Work from the outside of the frame.
4. To avoid cuts, remove the broken glass with a pair of pliers.
5. Remove old putty and glazier points using a putty knife and a pair of pliers.
6. Place a thin ribbon of putty in frame, pressing edges of glass firmly against the putty.
7. Place the glazier points carefully into the frame, against the glass. Avoid glass damage. Place the glazier points near the corners first then about every six inches along the glass.
8. Fill the groove with glazing compound or putty, pressing it firmly against the glass with a putty knife. Smooth the surface.

## Glass Cutting

If considerable glass cutting is required over a period of time, some equipment must be provided for this purpose. Glass-cutting equipment generally consists of a table about four feet square covered with a stretched blanket, a common T-square, and a glass cutter. One or two edges of the table can be marked off in inches. In some shops and hardware stores, a yardstick or rule is fastened to one edge of the table or cutting rack. Normal practice is to mark the location of the desired cut on the glass and then to draw a sharp glass cutter along the line. Only one pass with the cutter should be made; otherwise the glass will not always break along the desired line.

If cut pieces of glass are to be used for shelves or other similar locations, the edges can be rounded or smoothed by rubbing with a carborundum stone that has been dipped in water. In large glass-cutting shops, a special electric-powered grinding wheel, with the bottom section revolving through a bed of water, is used for smoothing the edges of cut glass.

The normal allowance for breakage in cutting and setting where considerable glass work is processed is about 5 per cent. Upon the completion of glazing work, all glass surfaces should be thoroughly cleaned, with all labels, paint spots, putty, and other defacements removed.

## Care of Windows

It is not recommended that ammonia or acid solutions be used to clean windows. To remove paint or putty smears, a cloth dipped in mineral spirits is generally used. Washing should be done with a mild detergent and water solution. It is good practice to establish a regular schedule for cleaning windows. The schedule should include removal of dirt and scale from window sash and frames, inspection for loose or crumbling putty, wire brushing of rust spots on metal sash and frames, spot painting these areas, and the calking of any open joints in window frames, sills, and jambs.

## CALKING

The use of calking material is necessary in construction to prevent water and air leakage at door and window frames, at flashings, at cracks between the siding and foundations, and in general at locations where materials meet without making a good seal. The calking of joints to prevent air and water leaks means a savings in air-conditioning and heating costs, and also prevents interior water stains and wood failure through rotting.

### Colors and Packaging

Calking compounds are available in at least 15 shades and colors, with the predomenant colors in use being white, gray, black, and aluminum. The material can be purchased in bulk or in cartridges. When purchased by the gallon (231 cubic inches per gallon), the weight is approximately 12 pounds to the gallon. Cartridges are preferred as they can easily be carried about and used with a minimum of waste. In addition, the use of cartridges prevents the possibility of thinning the material which could result in poor quality work.

### Calking Grades

There are two grades (1 and 2) of calking compounds that are generally used in the construction industry, and both are suitable for application at temperatures above 40°F. *Grade 1* is of soft consistency and is generally applied with a calking gun. This type is recommended for sealing joints where there is considerable contraction and expansion resulting from wide temperature variations. Gun application is rapid and economical, and is in general use. *Grade 2* is more like a glazing putty and is generally applied with a knife or puttying tool. It is often used at locations where a neat appearance is desired, particularly where the finished work is easily visible at close range.

## Sealers

On certain materials, such as cinder block, soft brick, and other porous building items, it is recommended that a primer or sealer be used. This prevents the loss of oil from the calking compound and maintains a good calking material consistency. It is suggested that a primer paint be applied to bare wood surfaces before using a calking compound. Often, linseed oil is used with good results.

Knot sealer is generally used when quick sealer drying is required. This type of sealing material sets in less than half an hour. Quick drying shellac and thinned varnish, as well as materials that dry to a hard, glazed surface, are not recommended to be used for sealing purposes, as they may prevent a good bond of the calking material.

## Work Preparation

It is essential that all areas to which calking material is to be applied be free of grease, rust, and dirt. Surface areas must be brushed clean and all old calking completely removed. Calking material does not adhere well to wet surfaces; therefore, cracks or joints where calking is to be applied should be dried prior to applying the material.

In order to save calking material, it is best to fill deep cracks with strands of oakum, calking cotton, or other similar material that is free from oil, to a point approximately 3/4-inch from the surface of the crack. In masonry materials, joints should be cut 1/4" wide, or to a width conforming to the job requirements.

## Calking Application

Although calking compounds are available in bulk lots, most of the material is purchased in tubes 10" to 12" long and approximately 2" in diameter. These tubes are used with a pistol-grip type of calking gun. The plastic ends of the tubes are cut to the required size.

If bulk materials are being used, it is recommended that a variety of gun tip sizes be obtained so that the proper tip can be selected for the size joint being calked. Cracks in gun-applied calking material can often be avoided by the use of the proper gun tip or correctly cutting the plastic end of the commercial calking tube. If the tip of the calking tube is not small enough to reach the bottom of the joint to be filled, an air gap may be left which may result in cracks in the finished work. The material should be forced into the grooves with sufficient pressure to expel all air and fill the groove solidly. The calking should be free of wrinkles and uniformly smooth when complete.

After the calking material is applied, the compound should be flush with the wall surface. A beading tool is often used to provide a desired finished appearance and also to pack the compound firmly into the joint particularly on masonry work. Prior to the start of calking, all dust and other debris should be removed from the grooves to be calked. After completing the work, clean all smears and other soiling from the adjoining surfaces to provide a neat and clean finished job.

CHAPTER 7

# Sheet Metal

Sheet metal is used extensively in conjunction with construction, building, and industrial work. It is used for roof gutters and roofing; heating, ventilating, and air-conditioning ducts; for machine guards and dust collection hoppers; and a variety of other purposes including tool boxes and kitchen canopies. Not all sheet metal is alike and each has its own characteristics and particular uses.

## COPPER

Copper sheets are fabricated into many building use items because of its great resistance to corrosion. Copper sheets are easily fabricated, brazed, and soldered, and are used extensively for gutters and gutter conductor pipes, roofing, and flashings. Many items requiring a resistance to corrosion are fabricated from copper, including screening. Lead-coated copper is installed in some locations to prevent the copper itself from staining the sides of structures.

Copper sheets are classified into two groups. *Soft-rolled copper* is generally used for standing-seam roofing and thru-wall flashings. Where a degree of stiffness is required, *cold-rolled* or *hard-rolled copper* is used, particularly for such purposes as gutters, downspouts, and base and counter flashings.

Copper is considered a permanent metal for building use and little maintenance is required. However, periodic inspections must be made to check for enlarged holes around screws, rivets, and nails. Seams occasionally crack and the solder may break or loosen.

Heavy wind or hail storms may damage copper, and stones thrown onto copper roofing may puncture the material. Prompt repair of all defects should be made.

## GALVANIZED IRON

Galvanized iron is steel or iron coated with zinc for the purpose of creating a corrosion-resistant material. It is medium priced and often used for ducts, gutters, machine guards, and similar metal-fabricated items. Common gauges in use are 24 and 26, although heavier and lighter gauges are available.

Fabricated seams are generally used in conjunction with this type of metal. Where tight connections are necessary, solder is applied. The material is not normally welded or brazed, as the process destroys the zinc coating.

Paint is a good protective coating for galvanized-iron products, but should not be applied to new material unless the surface has been treated. Cleaning with soap and water and wiping with turpentine or mineral spirits will clean the factory grime and grease that may have adhered to the sheets during manufacture. If the sheets have not weathered for at least six months, a mixture of household vinegar and water will properly prepare the surfaces to receive paint. The surfaces should be washed with clean water after the vinegar solution application.

When the zinc coating of galvanized iron is damaged or eroded, painting is necessary to prevent corrosion. Painting cycles for galvanized iron vary from 4 to 7 years, depending on specific conditions.

## ALUMINUM

Aluminum sheets are less rigid and do not have the strength of steel, but are lighter, have good appearance and, under average conditions, are more corrosion resistant than galvanized iron. Aluminum sheets are used for ducts, ventilators, gutters, screens, louvers, and other fabricated items in building construction. Most aluminum in use is 0.032" thick.

When installing aluminum, it is recommended that direct contact with mortar, cement, lime, and wood be avoided in order to prevent possible corrosive action. Aluminum surfaces in contact with these materials should be coated with a heavy-bodied bituminous paint or liquid asphalt.

Aluminum requires little maintenance except periodic inspections for physical damage, possible corrosion, and expansion and contraction damage.

## BRASS AND BRONZE

Brass and bronze are both copper-alloy metals and are used for interior and exterior metalwork. Although both are associated closely with name plates, bronze has a wide variety of uses and can be obtained in a variety of shapes, both extruded and drawn. The material is used in construction and for statuary. The utilization of brass and bronze in construction often serves a dual purpose of being functional as well as ornamental, such as its use on store fronts and elevator enclosures. The material has a long-term low-maintenance cost, needing nothing more than periodic polishing in certain applications.

The green deposit (oxidation) which appears on copper, brass, and bronze can often be removed by cleaning with a cloth dipped in ammonia. Vinegar into which salt has been dissolved will remove most tarnish and corrosion, but the surfaces should be washed with water after cleaning to prevent the vinegar solution from pitting the metal.

When new bright finishes are desired, all old lacquer and grime may be removed with the application of a paint thinner followed by a scrubbing with steel wool or pumice and water. Tarnish can be safely removed from bronze by swabbing with a concentrated solution of oxalic acid. Mineral acids are not recommended. To obtain a uniform surface color, it may be necessary to firmly rub the surface with steel wool. After the metal has been cleaned and it is dry, a new protective coating can be applied. Colorless lacquers are usually used for this purpose. Clear fingernail polish can be used for small objects.

If brass or bronze lacquered surfaces are in good condition, they should be cleaned about every two weeks and a light coating of lemon oil, paraffin oil, or wax rubbed on. Brass polishes that are satisfactory for this purpose are available.

## TIN PLATE

When sheets of iron are coated with a thin layer of tin, the material is usually referred to as *bright tin*. When it is coated with a tin-lead alloy, the term used is *terneplate*. Terneplate is usually a coating of 75 per cent lead and 25 per cent tin on sheet steel. When reference is made to 20- or 30-pound coatings, it means that each 100 pounds of sheet steel has been coated with 20 or 30 pounds of the alloy. Tin-plate material is normally used for roofing and flashing purposes, although copper, aluminum, or galvanized sheet steel is more popular.

## STAINLESS STEEL

Two great advantages of stainless steel are its durability and appearance. Made from alloys of steel and chromium (and sometimes nickel), with or without additives of silica, the material is used in locations where strength, corrosion and acid resistance, as well as beauty, are desirable. It can be formed and welded.

Care must be used in selecting the finish as it affects the appearance and price of the completed product. Cold-rolled mill finishes are normally designated as 2B and 2D types. Polished finishes range from No. 3 through No. 7, with the No. 3 type having a bright, grained appearance, and the most highly polished No. 7 type having a highly-reflective mirror-like surface. Naturally, the better the finish and reflectiveness, the higher the cost of the material.

Cleaning stainless steel can be acomplished with a good cleaner, but experience has indicated that cleaners containing chlorides have a tendency to cause pitting if not removed completely. Paste wax has been used with good success in polishing and maintaining interior surface. Outdoor appearances can be maintained by washing occasionally with ordinary soap and water.

## SOLDERING

Good soldering practices require a great degree of material cleanliness before and after the work is accomplished. Joints to be soldered must be free of foreign matter, such as corrosion, grease, paint, and dirt, to create a good bond. After soldering, the work must be cleaned in order to prevent staining and possible corrosion by the flux.

The most frequently used solder is half-and-half (50-50) which is composed of 50 per cent tin and 50 per cent lead. Where increased strength is required, 75-25 is used (75 per cent tin and 25 per cent lead). Silver solder is used where maximum strength of the soldered joint is required.

Aluminum solders are classified as hard solders, and a blow torch is normally used when soldering. This type of solder is an alloy of zinc, tin, and a small amount of aluminum.

### Flux

A soldering flux is used to complete the bond between the solder and the metal. Pure or diluted muriatic acid is commonly used, with the pure acid being used for galvanized iron, and diluted muriatic acid for soldering copper, brass, pewter, and terneplate. Diluted muriatic acid is made by adding small pieces of zinc or galvanized iron to the pure acid until the solution reaches a saturation point.

Tallow is often used for soldering lead, and it has been used successfully for soldering aluminum, although special fluxes are manufactured for this purpose.

Many good commercial soldering pastes and fluids are manufactured for soldering and are available at most hardware stores and lumber yards.

### Soldering Irons

Soldering irons are available in a variety of shapes and sizes. Small, light work requires a small iron; heavy work, including soldering of roofing seams, requires a heavy iron with a blunt point. The copper-type, with an iron shank and wooden handle, is heated in a gas or electric furnace, or with an acetylene or butane torch. Gasoline blow torches are also used for heating soldering irons.

When soldering, it is essential that the point of the iron be properly tinned. To do this, first file the point until it is bright. Heat the point and coat it with solder, using a bar of sal ammoniac, if necessary, to make the solder stick to the point. It is necessary to draw the copper iron slowly along the work to allow complete heat penetration to the metal being soldered. Do not disturb the solder until it has hardened. Lack of heat penetration will result in a poor job.

## WELDING AND BRAZING

Brazing is a process of joining two or more pieces of metal together with an alloy metal usually composed of copper or copper alloy. The composition of the filler rod, however, may vary. Brazing is often used for nonferrous metal as it forms a joint of high strength.

Copper-alloy brazing rods are used for joining steel, cast iron, brass, iron, and bronze. The flux used for these metals is a borax compound. Aluminum-alloy rods are used for brazing aluminum. Silver-copper rods, with phosphoric acid for a flux, are used when brazing stainless steel.

As in soldering, the surfaces to be brazed must be clean, properly aligned, and clamped prior to the start of actual brazing. The work is usually accomplished with an oxy-acetylene torch, with the brazing rod first being heated and dipped into the flux, and then fed to the joint as the torch is applied.

## SHEET-METAL SCREWS

Sheet-metal screws (sometimes called self-tapping screws) are used to hold joints and sheets together where extra holding power is required or where the assembled sections or components must be periodically dismantled. Holes for this type of screw are drilled or punched a little smaller than the screw diameter; the screw cuts its own thread as it is screwed into the opening. Sheet-metal screws are available in a variety of sizes and types, the most common being the round, pan, and flat head, with either Phillips or plain slot heads. These various types of sheet metal screws are shown in Fig. 1.

## RIVETS

If brazing or welding is not practical, and strong joints are still necessary, rivets are used as fasteners. Many varieties are avail-

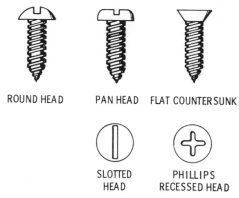

ROUND HEAD    PAN HEAD  FLAT COUNTERSUNK

SLOTTED HEAD    PHILLIPS RECESSED HEAD

**Fig. 1. Sheet-metal screws.**

able, including round head, flat head, mushroom head, and tinner's rivets. Tinner's rivets (coated type) are used if soldering is required after riveting; flat-head types are used for light-gauge metals; and round-head types for heavier metals where extra strength is necessary.

Rivet size is normally determined by the weight per thousand. As an example, 3-pound rivets weigh 3 pounds for each 1,000 rivets. The size for large rivets is given by the length and actual diameter.

When placing rivets, holes should be punched or drilled so as to permit a snug fit. After inserting into the hole, the rivet is backed up by an iron block, the indention in a rivet setting tool is placed over it, and the tool is struck a sharp blow with a hammer to bring the two pieces of metal to be joined together. The rivet is then flattened with the hammer, after which the cup of the rivet set is placed over the flattened rivet and the tool is again struck sharply with the hammer to form a rivet head.

## FLASHINGS

Various metals are used to prevent the seepage of water at critical points of a structure. Particularly vulnerable are those locations where sloping roofs and walls join. There are many types of flashing uses and applications.

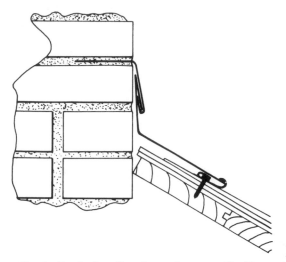

**Fig. 2. Typical wall and metal counter flashing.**

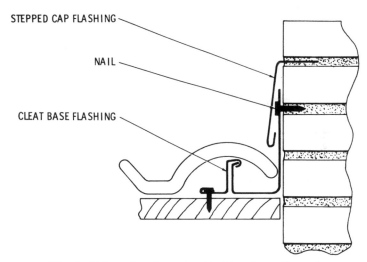

STEPPED CAP FLASHING

NAIL

CLEAT BASE FLASHING

**Fig. 3. Sidewall and counter flashing for a tile roof.**

## Wall Flashings

Wall flashings do not always have a cap flashing of metal, particularly if the base flashing is protected. In most instances,

213

however, sidewall cap and base flashings are used at the intersection of the roof and vertical walls. See Figs. 2 and 3 for these types of applications.

## Thru-Wall Flashings

Thru-wall flashings (Fig. 4) are normally installed during construction of the wall at locations where leakage may occur. Special corrugations in commercial thru-wall flashing provides a mechanical bond with the mortar bed. It is good practice to apply a thin layer of mortar both over and under the metal, sloping the metal in the direction of the water drain.

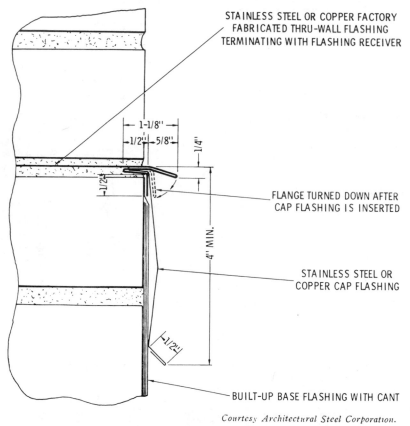

*Courtesy Architectural Steel Corporation.*

**Fig. 4. Thru-wall flashing.**

## Ridge and Hip Flashings

Ridge and hip flashings are usually fabricated in 8-foot lengths and installed with a 3-inch overlap. All installations must be watertight. Typical ridge and hip flashings are shown in Fig. 5.

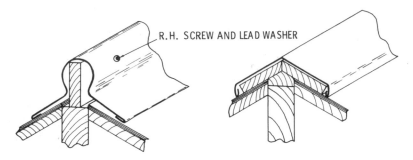

R.H. SCREW AND LEAD WASHER

*Fig. 5. Ridge and hip flashings.*

Flashings can be fastened with a slide-type prepared fastener, round-head wood screws and lead washers (which may or may not be soldered), or with fastening cleats.

## Valley Flashings

The most common valley flashing is the open type (Fig. 6) which is laid in an 8-foot continuous sheet with overlapping seams. Each side extends a minimum of 5 inches (6 inches is standard practice) under the roofing for the entire length. The sides of the sheets are fastened on 12-inch centers with cleats. If the

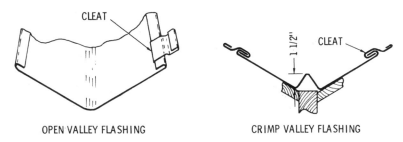

CLEAT

CLEAT

1 1/2"

OPEN VALLEY FLASHING          CRIMP VALLEY FLASHING

*Fig. 6. Valley flashings.*

215

roof slope is steep, a V-crimp is placed in the center as a baffle to prevent water from washing over the valley and up under the roofing on the other side.

Closed valley flashings (Fig. 7) are generally used with shingle roofs where the slope is more than 8 inches to the foot. This

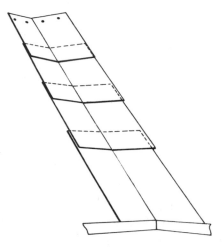

**Fig. 7. Closed valley flashing.**

type of flashing is not recommended for roofs having an average or medium slope. The sides of the flashing should be placed far enough up under the shingles so as to prevent any water from backing up under the roof.

## Chimney Saddles

Cap and base flashings are normally used for most chimneys, but those chimneys which go through a sloped roof also require a saddle on the slope side to prevent water and snow from settling in this pocket. Chimney saddles (or *crickets*, as they are sometimes called) are made of a variety of metals, depending on specific conditions. Saddle flashings should be carried under the shingles for at least 6 inches and the folded edges fastened with cleats. A typical chimney flashing is shown in Fig. 8 and a typical saddle in Fig. 9.

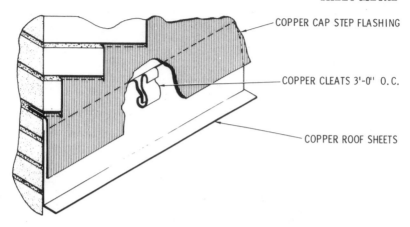

COPPER CAP STEP FLASHING

COPPER CLEATS 3'-0'' O.C.

COPPER ROOF SHEETS

*Fig. 8. Chimney cap and base flashing.*

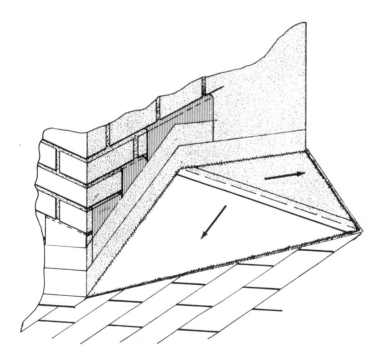

*Fig. 9. Chimney saddle.*

## Gravel Stops

Gravel stops are installed at the edges of roofs and around drains when built-up roofs are installed on buildings. Gravel stops less than 30 feet long are formed with lapped and soldered joints. For those exceeding 30 feet, a loose lock expansion joint is provided. Fig. 10 illustrates typical gravel stops.

The rib-bond type facia and/or gravel stop in Fig. 10 is of the embossed herringbone type which has a tendency to decrease buckling and waviness due to temperature changes.

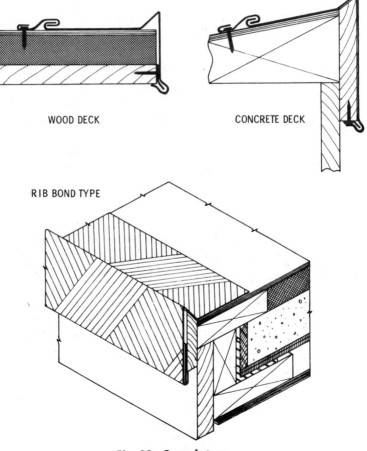

WOOD DECK                    CONCRETE DECK

RIB BOND TYPE

*Fig. 10. Gravel stops.*

## METAL ROOFS

The most common type of metal roofing is the standing-seam type. Sheets should not be larger than 20 × 96 inches; the finished seam should be at least 1 inch high, and the distance between seams should not be greater than 17 inches. Unless condi-

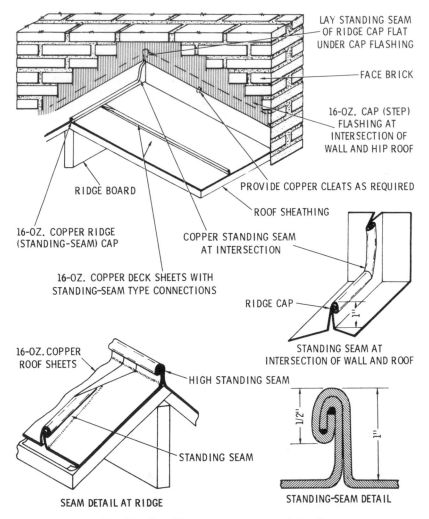

LAY STANDING SEAM OF RIDGE CAP FLAT UNDER CAP FLASHING

FACE BRICK

16-OZ. CAP (STEP) FLASHING AT INTERSECTION OF WALL AND HIP ROOF

PROVIDE COPPER CLEATS AS REQUIRED

ROOF SHEATHING

RIDGE BOARD

16-OZ. COPPER RIDGE (STANDING-SEAM) CAP

COPPER STANDING SEAM AT INTERSECTION

16-OZ. COPPER DECK SHEETS WITH STANDING-SEAM TYPE CONNECTIONS

RIDGE CAP

STANDING SEAM AT INTERSECTION OF WALL AND ROOF

16-OZ. COPPER ROOF SHEETS

HIGH STANDING SEAM

STANDING SEAM

SEAM DETAIL AT RIDGE

STANDING-SEAM DETAIL

**Fig. 11. Standing-seam copper-roof details.**

219

tions dictate otherwise, standing seams are not soldered, thus permitting lateral movement. The greatest majority of standing-seam roofs are fabricated of copper. Fig. 11 shows the details of a standing-seam roof.

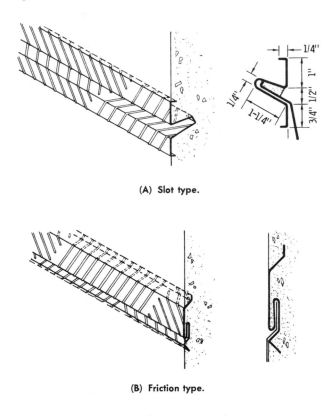

(A) Slot type.

(B) Friction type.

*Fig. 12. Sheet-metal reglets.*

## REGLETS

When sheet metal is secured to concrete or masonry and water penetration is not desired, reglets are often used. These are installed before placing the concrete by nailing them to wooden forms. When the forms are removed, the reglet is firmly installed and the flashing may be secured to it with metal plugs or lead-wool packing. Fig. 12 shows two types of reglets.

## ROOF GUTTERS

Gutter materials vary, but most are fabricated from galvanized iron, copper, or aluminum. Sizes vary depending on requirements, but installation of the half-round type in sizes less than 4 inches is not recommended. The recommended slope of the gutter to the gutter outlet is $1/16''$ per foot.

Expansion joints are required approximately every 30 feet for built-in gutters and at intervals ranging from 20 to 40 feet for hanging gutters, particularly if they are made of copper. Care must be used to allow the gutter ends a minimum clearance of at least $1/2''$ from vertical walls to allow for expansion.

Many gutter shapes and types are available, but the most commonly used are the box and ogee type, and the half-round single-bead type. These types are illustrated in Fig. 13.

Leaf screens are used in downspout gutter outlets in areas where many trees exist to prevent the leaves from clogging the downspout. The screen material should be the same as the gutter material. A copper screen should be used with copper gutters and galvanized screen with galvanized gutters.

In order to avoid water flowing around sharp bends, gutter outlets should be placed at corners. The recommended spacing of gutter outlets is 40 feet, although this distance may vary depending on specific job conditions and downspout sizes.

## DOWNSPOUTS

Downspouts (sometimes called conductors or leader pipes) are used to carry rainfall from the gutters to the ground for runoff, or to storm sewers. Almost all downspouts are of a corrugated design, as this type is less likely to be damaged by freezing than the plain metal type. The corrugations offer good resistance to freezing and thawing and possible failure. The downspout should have as few bends as possible, as these gather debris which leads to pipe failure. The pipe is normally supplied in 10-foot lengths and is installed with lapped and soldered joints. Sheet-metal screws are applied for additional strength. If copper piping is installed, slip joints at 20-foot intervals are recommended to allow for contraction and expansion.

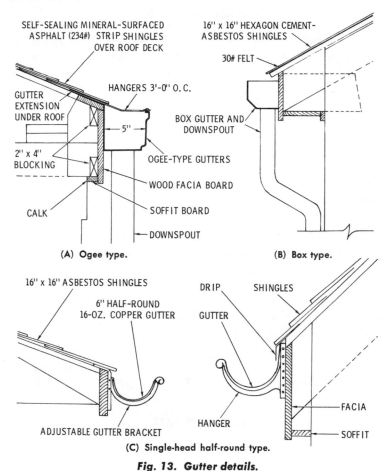

**Fig. 13. Gutter details.**

It is essential that inspection of all gutters, downspouts, and flashing be conducted at least twice a year. Spring and fall cycles are recommended for inspections, at which time all debris and other foreign matter can be removed.

All flashing should be inspected to assure water-tight connections, possible corrosion, and the condition of calking and cap-flashing fastenings. Gutters and downspout must be inspected for possible stoppages and buckling, and for loose straps and hangers. Painted metal work should be checked for paint failure.

Some metal work, such as roof blower extensions and smoke pipe, may have guy wires which should be inspected for the condition of the guys and fastenings. Repair as necessary. Gutters must be cleaned and the interior of the galvanized type should be inspected for rust spots. Liquid asphalt and red-lead base paint are both good materials for interior coating of galvanized iron gutters if the gutter has weathered for at least six months. The soldered joints of copper gutters should be checked for failure, and resoldered if necessary.

Repairs and maintenance are less expensive if a good periodic inspection program is maintained, as possible serious failures may be corrected in time to prevent excessive damage and repairs.

## DUCT WORK

Sheet metal is used extensively in fabricating duct work for the purposes of transferring air and fumes in buildings. Good design requires that the interior of ducts be as smooth as possible to prevent obstruction and trubulence in the air flow. Sharp turns should be avoided in order to reduce pressure losses, and transition-type fittings should be made as long as possible to create a better air flow. Tests have indicated that good elbow design in flat or rectangular duct work reduces air-pressure loss if the centerline radius is 1-1/4 times the width of the elbow in the plane of the turn. If an elbow does require a small inner radius, the loss of air pressure can be reduced by the installation of turning vanes (Fig. 14). Turning vanes also work well in square throat elbows.

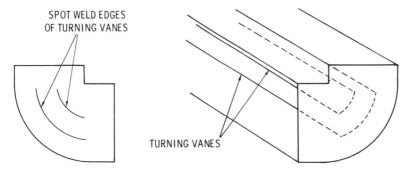

*Fig. 14. Turning vanes in square-throat elbows.*

The two most common types of seams used in fabricating duct work are the Pittsburgh and grooved types, as in Fig. 15. Both are used to join two metal edges together. Most ducts are assembled with the use of the Pittsburgh-seam method. This type can be made on a bending brake, but a special Pittsburgh-seam machine which quickly forms the seams on almost all lightweight metals is usually used.

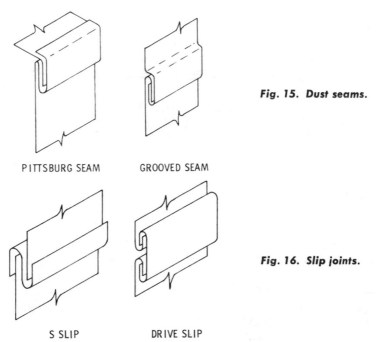

**Fig. 15. Dust seams.**

PITTSBURG SEAM          GROOVED SEAM

**Fig. 16. Slip joints.**

S SLIP          DRIVE SLIP

Continuous lengths of finished ducts are joined together with S or drive slips (Fig. 16), with the drive slip being applied to the long side of the duct. Drive slips are applied at the two sides, driven into place, and then locked by bending over the ends to close the corners.

In order to avoid vibration noises in a duct system, connections at outlet locations are usually made with a heavy canvas connection. Asbestos fiber material is used if the possibility of a fire hazard exists. The cloth or canvas is folded with the metal, and reinforced at either end of the connection by inserting screws through both the metal and cloth, as shown in Fig. 17. If the duct is wider than 12 inches, additional reinforcement screws near the middle of the connection may be required.

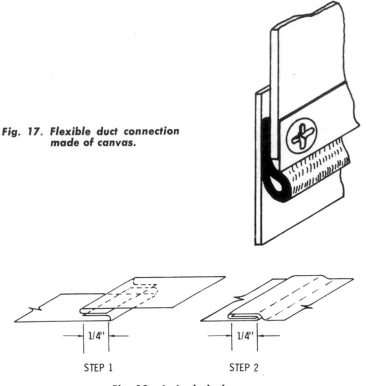

**Fig. 17. Flexible duct connection made of canvas.**

1/4"          1/4"

STEP 1          STEP 2

**Fig. 18. A single lock seam.**

Single lock seams are used with almost all lightweight metal work. These seams are easily made on a brake or folding machine (Fig. 18). Finishing is completed with a hand groover and a hammer.

## SHEET-METAL PATTERN DEVELOPMENT

Three general methods are used to develop sheet-metal patterns—the *parallel line*, the *radial line*, and the *triangulation* method.

225

## Parallel Line Development

The parallel line method is most commonly used and includes the development of box patterns and all patterns where the true line lengths in a three-view drawing are known or indicated. In a box pattern, the bottom size is laid on the pattern material and the sides and ends added, as indicated in Fig. 19.

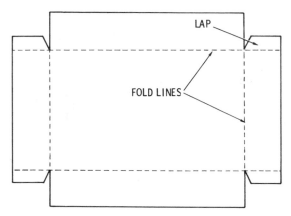

*Fig. 19. A typical box pattern.*

In a round-pipe "T" intersection, the following procedure is used: Lay out the true size front view, as in sketch 1 in Fig. 20, strike off a half construction circle and divide it into 6 equal parts by placing a radius-set compass at points 1, 4, and 7, and swinging an arc across the half circle. It now has been divided

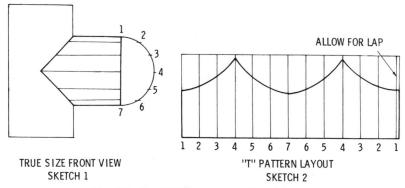

TRUE SIZE FRONT VIEW
SKETCH 1

"T" PATTERN LAYOUT
SKETCH 2

*Fig. 20. Parallel-line pattern development.*

into 6 equal parts. As the half circle is a half construction pattern of 6 equal parts, the 6 equal sections of the circle are doubled to 12 (for a full circle) and laid out as indicated in sketch 2. The true lengths of the intersecting pipe are obtained from the true size front view, laid out on sketch 2 as indicated, and the intersection "T" pattern is ready to be cut out. Lap allowance must be added.

After the intersecting "T" section is formed and seamed or riveted, it is positioned on the pipe, the required opening traced with pencil or crayon, and then cut with an allowance of approximately 1/2 inch for an interior lap for peening purposes. If the material is heavier than 26 gauge, the interior lap section must be cut at approximately 3/4-inch intervals up to the traced required opening edge of the pipe, making it easier to bend the lap to the interior of the intersection "T" pipe section.

## Radial Line Pattern Development

A front-view plan is made of the required cone, and a half-plan circle developed on the lower portion, as in Fig. 21. This

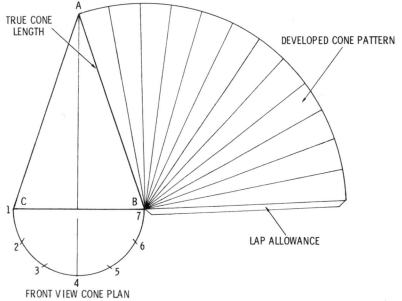

*Fig. 21. Radial-line pattern development.*

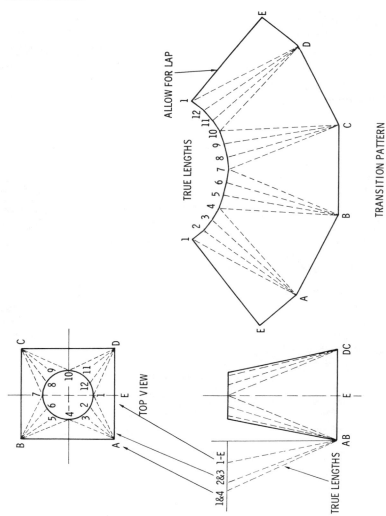

**Fig. 22. Triangulation pattern development.**

half plan is divided into 6 equal parts by striking off the half plan with a compass which was used to develop it. The compass or divider must be held at the radius set (half the diameter) and placed on points 1, 4, and 7 of the half-circle development. Since the true length of the cone is known and indicated in the front-view plan of the cone, the compass is set on this true length at point 1 and point *a* (*a* is the apex of the cone), and the pattern arc is drawn. This true-length arc is then divided into 12 equal parts (the space from point 1 to 2). The 12 parts equal the full circumference of the bottom part of the cone. After the arc has been divided, lines are drawn to point *B*. An allowance is made for the lap, after which the pattern may be cut out.

## Triangulation Development

A triangulation pattern layout is used in the development of transition pieces for ducts and other sheet-metal items to make connections for different size and shaped pieces. Fig. 22 shows a method of general pattern development for a square-to-round transition.

The top view round opening of the transition is divided into 12 equal parts, as indicated in the sketch. The segment section points are connected to the corner parts of the square, points *A*, *B*, *C*, *D*, and also to *E*.

A front-view layout of the pattern is made, and adjacent to this a horizontal true-length line is established and true-length distances established. Following Fig. 22, the transition pattern can be developed. Allow a lap for seaming, riveting, or welding.

Use the true-line markings on the pattern to form the transition in a metal brake. When a metal brake is not available, a railroad rail or a 2″ × 4″ or 4″ × 4″ piece of lumber may be used for forming the pattern.

CHAPTER 8

# Electrical Maintenance

A general recommendation from electricians and safety experts is to be extremely careful when working with electrical circuits and electrical power sources. Many persons have been injured and some killed through careless handling of appliances, circuits, and power-source conductors.

In general, direct electrical current can be compared to water and its flow. Even as water flows through a pipe, electrical current flows through electrical wire. In a water system, pressure is measured in pounds. In an electrical system, the measure is in *volts*. In both, the quantity of flow will vary with the pressure or supply. In the study of water (hydraulics), a gallon is a specific quantity and in electricity the *ampere* is a specific unit of measure. However, neither is the measure of the rate at which it flows or is delivered. The normal term of measurement in water (hydraulics) is in gallons or cubic feet per minute and in electricity, *amperes per second*. The flow of water in a water pipe is lessened by internal friction and is measured in pounds per foot. The flow of electrical current is governed in an electrical wire by the amount of resistance it offers and the amount of resistance varies in accordance with the wire material. This resistance is measured in *ohms*. As in the water pipe, the longer the run the greater the resistance. Using a larger size pipe will reduce friction in a water system and if a larger size wire is used in an electrical system, resistance is reduced. However, at this point the comparison of water and electricity flow ends.

In practice, there is a big difference when pressure on a water pipe increases and pressure (voltage) on an electrical wire in-

230

creases. In a water pipe, friction losses are lowest when water moves through the pipe slowly. There is always a specific relationship between resistance (ohms), pressure (volts), and current (amperes) when electrical current flows through a wire of a known length. The common formula being *amperes equals voltage divided by resistance.* Voltage increases result in greater delivery of electrical current over a smaller wire.

Thus, when electricity is transmitted over electrical lines for distribution, the voltage is increased over the line and reduced at the point of use by "stepping it down" with transformers for safe use.

## ELECTRICAL UNITS OF MEASURE

In order to arrive at units of electrical measure, the *volt, ampere, ohm,* and *watt* have been adopted and these units are commonly used throughout the world.

The *volt* corresponds to pressure and is the *unit of measurement for electric pressure.*

The *ampere* is a given or certain amount of electricity and can be defined as the practical *unit of electric current flow.* Just as water passing through a pipe is defined as "x number of gallons per second" thus electricity is measured by "x number of amperes per second" flowing through a conductor.

The *watt* can be termed the unit of work or energy. It is the unit which shows current consumption with both voltage and amperage considered. For example:

1 ampere at a pressure of 1 volt = 1 watt
1 watt used for 1 hour = 1 watt hour
1000 watt hours = 1 kilowatt hour (kWh), which is the unit by which electricity is metered.

A list for estimating the amount of current (watts) required by various household and farm appliances is as follows:

Stereo Hi-fi ................................................................300 watts
Electric Shaver ........................................................10 watts

Room Heater ...............................................................1600 watts
Ceiling Light ...............................................................100 watts
Sunlamp ...............................................................275 watts
Single-Tube Fluorescent Fixture ...............................50 watts
Table Lamp ...............................................................100 watts
Television Set ...............................................................300 watts
Vacuum Cleaner ...............................................................400 watts
Coffeemaker ...............................................................600 watts
Refrigerator ...............................................................250 watts
Food Mixer ...............................................................150 watts
Deep Fryer ...............................................................1300 watts
Rotisserie ...............................................................1400 watts
Large Roaster ...............................................................1380 watts
Automatic Toaster ...............................................................1100 watts
Ironer ...............................................................1650 watts
Hand Iron ...............................................................1000 watts
Sump Pump ...............................................................300 watts
Fuel-fired Furnace (oil) ...............................................800 watts
Small Drill Press ...............................................................500 watts
Electric Range ...............................................8000 to 16,000 watts
Hot Water Heater ...............................................................2500 watts
Washer and Dryer ...............................................................5500 watts
Dishwasher ...............................................................1800 watts
Garbage Disposer ...............................................................900 watts
Farm Milk Cooler ...............................................................450 watts
Farm Cream Separator ...............................................300 watts
Farm Milking Machine ...............................................400 watts
Water Pump ...............................................................300 watts
Chick Brooder (Farm) ...............................................1000 watts
Yard Light ...............................................................150 watts

In order to convert wattage into amperage, multiply watts by 0.0087. As an example, a water pump of 300 watts, a chick brooder of 1,000 watts, and a yard light of 150 watts add up to a total of 1450 watts. Multiplied by 0.0087, the result is 12.61 amperes.

The unit of internal force (resistance) which holds back current flow in a conductor is called the *ohm*.

In electrical terminology, the *kilowatt* is the term used to express a thousand watts. One horsepower equals 746 watts.

The *kilowatt-hour* is the total amount of electrical power used in an hour when the flow rate remains constant at 1 kilowatt.

## DIRECT CURRENT

Direct current means that the electrical current travels in one direction only. Current from batteries is the direct type. However, electric energy used in ordinary power and lighting could be either alternating or direct.

## ALTERNATING CURRENT

Alternating current flows in a given direction at a specific moment, dies to zero, and then flows in the opposite direction. It is normally three phase, but can be single phase, polyphase, or 2-phase. In effect, we are referring to a single-cylinder, 2-cylinder, or 3-cylinder engine. When alternating current makes a complete change per second, and when referring to 60 cycles, the current has made 60 complete changes or reversals per second.

## MAINTENANCE OF ELECTRICAL WIRING SYSTEMS

Maintenance functions are determined to a great extent by the location, selection, and installation of the original equipment and wiring. Operating breakdowns normally can be prevented by a good maintenance program of periodic inspections, tests, adjustments, cleaning, lubrication, and tightening of connections.

### Lighting Fixtures

Flashing fluorescent tubes must be replaced, as they result in continuous working of the starter and eventual failure of the lamp ballast. Keep the tubes clean and replace darkened tubes, as they are approaching the end of their life expectancy.

Lighting fixtures should be cleaned with a detergent or cleaning spray, as dry-wiping normally leaves dirt on the light reflector.

Trained maintenance men should replace burned-out incandescent bulbs, particularly in hard-to-reach locations, as replacement by inexperienced help with poor equipment may result in a fall and injury.

## General Inspection

Inspect for defective convenience outlets and switches, and for the improper use of extension cords. Inspect for any condition that may cause a fire and for nonstandard or unauthorized electrical attachments or appliances.

## Distribution Panels

Fuse or switch cabinet should always be accessible for service. Avoid placing furniture, boxes, or other items in a location that may interfere with ready access to the fuse or switch cabinet. Circuit identification must be up to date. Problem circuits, trouble areas, and fuse replacement is simplified if service boxes and circuits are properly identified.

## Wire (Conductor) Connections and Loose Fittings

All electrical connections of conductors, including splices, terminal connections, and wire taps, should be inspected at regular intervals scheduled to particular needs and situations. Partly contacting, loose, or damaged connections at terminals, or poor splices may result in arcing or short circuits and inefficient operation of the electrical system. These deficiencies should be corrected when discovered.

Fittings, such as conduit couplings, connectors, and box entry, should be inspected for looseness or separation. Tightening of circuit bushings or lock nuts may prevent future troubles particularly in areas of high vibration.

## Insulation

Electrical conductors (wires) are insulated during manufacture to insure protection against accidental contact with other materials which may conduct electricity. This coating of material over the wire prevents contact of the conductor with piping, outlet boxes, and other materials which could result in short circuiting of the system. All frayed or damaged insulation or loose connections should be taped or replaced to prevent possible trouble. All repairs should be made in a permanent manner as temporary work can be a fire hazard and eventually must be replaced.

## Extension Cords

Care must be used in handling extension cords in order to assure long life. It is improper to pull a cord from a wall plug by using the cord itself as a means of removal. Instead, grasp the plug to remove the cord.

Short circuits and blown fuses often result from damaged extension cords. Always repair damaged or frayed cords as they can be a fire hazard. If beyond economical repair, they should be destroyed. The Underwriter's knot shown in Fig. 1 should be used when making repairs to protect the cord from strain.

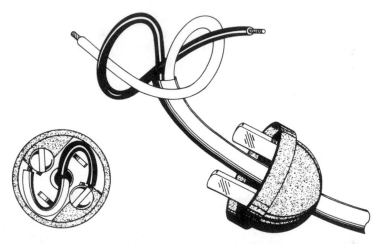

**Fig. 1. An Underwriter's knot.**

## ELECTRICAL CIRCUIT FAILURES

There are many reasons for circuit failures, but the four most common are *loose connections, overloaded circuits, short circuits*, and *improperly installed fuses*.

## Loose Connections

Occasionally the screw in the fuse socket may be loose and the fuse does not make proper connection. If the bottom of the fuse is burned or pitted, pull the main switch and then tighten the loose screw.

## Overloaded Circuits

Normally, there are two types of overloads in circuits. One of a temporary nature, where an electric motor as it starts draws about three times its normal running current use. In cases of this type, a time-delay fuse of the proper size should be used. These slow-blowing fuses are available under various trade-names, such as *Slo-Blo, Fusetron,* etc.

In case of a constant overload, some of the connected loads should be shifted to another circuit or to new circuits added to the system.

A test of the electrical circuit can be made to locate sources of overload with a testing device that can be purchased at most electrical stores. The testing device comes with a complete set of instructions for its use. The device is not expensive and locates current leakages in motors and appliances and finds defective cords and locates short circuits.

## Short Circuits

When a short circuit occurs, the blown fuse window will normally be discolored. In effect, it means that the conductor is in contact with another bare wire or metal surface, resulting in a "short."

To find the "short," disconnect all lamps and appliances. Turn off the main panel switch, replace the fuse, turn on the panel switch and then start plugging in the lamps, appliances or electrical devices that were on the circuit. When the faulty electrical device is plugged in, the fuse will "blow" again. Examine the cord, lamp, or appliance and, in all probability, the device will have a bare wire that is causing the short circuit.

## Improperly Installed Fuse

Occasionally, fuses are not long enough to make proper contact with the fuse-socket base, or the fuse is not properly screwed into the socket. Upon examination, if the fuse window is not discolored or the metal strip broken, firmly screw the fuse into the socket or replace it with a new fuse of the same rating, as the situation requires.

## POLARIZED (GROUNDED) DEVICES

Polarized (grounded) receptacles are recommended in conjunction with electrical installations wherever possible. These devices have two current-carrying contacts or wires and one grounding wire. Installation of this type of device is recommended for use with portable tools, extension lights, appliances, and plug-in type machinery. The devices are installed in workshops, garages, basements, kitchens, and yard outlets, guarding against accidental shock from current leakage because of exposed line wiring.

Adaptors for converting standard receptacles to accommodate the new attachment are available at most electrical stores. Instructions for conversion are usually furnished with the attachment.

## THIN-WALL CONDUIT INSTALLATION

Thin-wall conduit should be installed only with steel switch and outlet boxes and never with porcelain or *Bakelite* types. A conduit bender should be used to make all bends. Empty conduit and switch boxes should be completely installed before the insulated conductors wires are placed in the conduit. On exposed runs, support straps should be installed every 6-feet, and on concealed runs, every 10-feet. Standard practice normally dictates an allowance of 8 inches of insulated wire at each box for connections.

Conduit size selection can be based on the following recommendations, unless local codes dictate otherwise.

1/2″ conduit carries four No. 14 wires or three No. 12 wires.
3/4″ conduit carries four No. 10 or No. 12 insulated wires or three No. 8 wires.
1-1/4″ conduit carries four No. 6 wires or three each of No. 2, No. 3, or No. 4 wires.
1-1/2″ conduit carries three No. 1 wires.
2″ conduit carries four No. 1/0 or three No. 3/0 wires.

## MOTORS AND CONTROLS

Motors for industrial applications are almost always of the 3-phase, 60-hertz type, although some of the smaller fractional-horsepower units are single phase.

Some motors have features such as overload protectors and sealed bearings. These extra features increase the cost of the motor, but they are recommended. Sealed bearings normally do not require lubrication. However, motors with other types of bearings must be lubricated at periodic intervals. The instructions that accompany the motor will indicate how often lubrication is required.

Modern motors and their associated controls will give long service with the proper care. Particular attention should be given to newly placed motors to detect possible installation discrepancies, such as alignment, anchoring, and connections which may lead to early failure.

## Preventive Maintenance Inspections

Positive preventive maintenance inspections should be programmed and should include the following at monthly intervals.

1. *Brushes.* Brushes worn unevenly should be trued up or replaced. Pigtail connections should be checked and tightened if required. The brushes should be replaced if they have chipped toes or heels or have excessive heat cracks. Brushes should work freely; if they stick, the holder should be cleaned.

2. *Commutators.* Commutators must be free of grime and oil in order to assure positive commutation. Clean as required.

3. *Oil Seals and Oil Rings.* It is essential that the oil seals, plugs, and well covers are tight and that all oil caps are closed so that grime cannot get into the bearings. Inspect the oil rings to be sure they are free and turn with the shaft.

4. *General.* Check the temperature of the motor by hand. If the hand can be held on the motor with reasonable comfort, it can be assumed to be operating within its temperature rating.

Keep motors free from dirt and moisture. Examine the ventilation openings to see that they are not impeding the flow of air.

Note and correct any unusual noises from the motor when it is in operation. Make sure that motor alignment is correct and that belts are not frayed and loose. Poor alignment may result in excessive bearing wear and possible failure.

Starters and controls must be kept clean, dry, tight, and operating at the recommended voltage. All connections should be inspected and tightened if loose.

Inspect for loose or missing blades on squirrel-cage rotors and replace and/or repair as may be required.

Windings of motors should be free of dust. Remove the dust with a cloth or a hand bellows, making sure that the dust is wiped from the housings and slip rings.

Make sure that all electrical connections are in accordance with the National Electrical Code and all local codes.

A motor of less than 1/2 hp can generally be plugged into a 120-volt outlet, but it is good practice to provide a 240-volt service for motors of greater capacity. Wire should be of sufficient capacity to safely handle the current demands of the motor. For safety and convenience, install switches at easy to reach locations. Ordinary lamp cords should never be used for motors; extension cords should be the heavy-duty type with No. 14 or heavier wire.

Table 1 can be used for determining the approximate wire size to use for single-phase motors.

## WIRING SAFETY PRACTICES AND COLOR CODING

In all cases, wiring installations and repairs must conform to the rules of the National Electrical Code. The Electrical Code requires that black conductors should always be used as hot wires and that white insulation should indicate a grounded conductor. In many installations, colored conductors, such as blue, red, and orange, may also be used for the hot leads. Green insulation normally indicates the equipment ground wire. The terminal color on an electrical device for connecting the hot wire is normally brass-colored, whereas the silver-colored terminal of the device is used for the white or grounded wire.

A fuse or a circuit breaker protects wires from overloads. Never use a fuse or circuit breaker rated higher than the ampere capacity of the wire, and *never* substitute a wad of foil or a coin for a fuse. Do not handle energized electrical equipment, including fuse boxes, when standing on a damp surface or wet basement floor.

The National Electrical Code is the recognized authority on guidelines for safe electrical practices. Local codes may vary but

## Table 1. Approximate Wire Sizes For Single-Phase Motors

| HP | Volts | Approximate Starting Current (Amperes) | Approximate Full-Load Current (Amperes) | Wire Size | | | | |
|----|-------|------|------|----|----|----|----|----|
| | | | | Distance From Main Switch to Motor | | | | |
| | | | | 25 ft. | 75 ft. | 100 ft. | 150 ft. | 200 ft. |
| 1/4 | 120 | 18 | 5 | 14 | 14 | 12 | 10 | 10 |
| 1/2 | 120 | 18 | 5 | 14 | 14 | 12 | 10 | 8 |
| 1/2 | 240 | 12 | 3 | 14 | 14 | 14 | 14 | 12 |
| 1 | 240 | 16 | 5 | 14 | 14 | 14 | 14 | 12 |
| 1 1/2 | 240 | 20 | 8 | 14 | 14 | 14 | 12 | 10 |
| 2 | 240 | 28 | 12 | 14 | 14 | 12 | 10 | 10 |
| 5 | 240 | 70 | 22 | 10 | 10 | 8 | 8 | 6 |

are established for safe electrical installation and maintenance procedures.

## ELECTRICAL INSTALLATIONS AND MAINTENANCE

The National Electrical Code provides that all electrical outlets will be grounded. Grounded outlets; three prong type receivers, permit the connection of two wire cords and plugs as well as three wire plugs and cords. For homes, weatherproof outlets are desirable on the driveway side of the exterior of the building installed about four feet above grade.

### Convenience Outlets

Convenience outlets in homes should be spaced no more than 12 feet apart so that no point on the wall measured along the floor will be six feet from an outlet. Kitchen, laundry, pantry and dining room outlets should be divided equally between two or more 20-ampere branch circuits. It is recommended that each kitchen outlet be installed on an individual circuit though this will increase installation costs.

Additional outdoor weatherproof outlets may be required to serve particular family needs.

## Table 2. Converting Kilowatts to Horsepower

| kW | hp | kW | hp | kW | hp | kW | hp |
|----|------|----|--------|-----|---------|------|---------|
| 1  | 1.341 | 19 | 25.471 | 60  | 80.436  | 190  | 254.71  |
| 2  | 2.681 | 20 | 26.812 | 65  | 87.139  | 200  | 268.12  |
| 3  | 4.022 | 22 | 29.493 | 70  | 93.842  | 220  | 294.93  |
| 4  | 5.363 | 24 | 32.174 | 75  | 100.545 | 240  | 321.74  |
| 5  | 6.703 | 26 | 34.856 | 80  | 107.248 | 260  | 348.56  |
| 6  | 8.044 | 28 | 37.537 | 85  | 113.951 | 280  | 375.37  |
| 7  | 9.384 | 30 | 40.218 | 90  | 120.654 | 300  | 402.18  |
| 8  | 10.725 | 32 | 42.899 | 95  | 127.357 | 325  | 435.69  |
| 9  | 12.065 | 34 | 45.580 | 100 | 134.048 | 350  | 469.21  |
| 10 | 13.406 | 36 | 48.261 | 110 | 147.47  | 400  | 436.24  |
| 11 | 14.747 | 38 | 50.943 | 120 | 160.87  | 450  | 603.27  |
| 12 | 16.087 | 40 | 53.624 | 130 | 174.28  | 500  | 670.30  |
| 13 | 17.428 | 42 | 56.305 | 140 | 187.68  | 600  | 804.36  |
| 14 | 17.768 | 44 | 58.986 | 150 | 201.09  | 700  | 938.42  |
| 15 | 20.109 | 46 | 61.667 | 160 | 214.50  | 800  | 1072.48 |
| 16 | 21.450 | 48 | 64.349 | 170 | 227.90  | 900  | 1206.54 |
| 17 | 22.790 | 50 | 67.030 | 180 | 241.31  | 1000 | 1340.60 |
| 18 | 24.131 | 55 | 73.733 |     |         |      |         |

## Table 3. Converting Horsepower to Kilowatts

| hp | kW | hp | kW | hp | kW | hp | kW |
|----|------|----|--------|-----|--------|------|--------|
| 1  | .746 | 19 | 14.174 | 60  | 44.76  | 190  | 141.74 |
| 2  | 1.492 | 20 | 14.920 | 65  | 48.49  | 200  | 149.20 |
| 3  | 2.238 | 22 | 16.412 | 70  | 52.22  | 220  | 164.12 |
| 4  | 2.984 | 24 | 17.904 | 75  | 55.95  | 240  | 179.04 |
| 5  | 3.730 | 26 | 19.396 | 80  | 59.68  | 260  | 193.96 |
| 6  | 4.476 | 28 | 20.888 | 85  | 63.41  | 280  | 208.88 |
| 7  | 5.222 | 30 | 22.380 | 90  | 67.14  | 300  | 233.80 |
| 8  | 5.968 | 32 | 23.872 | 95  | 70.87  | 325  | 242.45 |
| 9  | 6.714 | 34 | 25.364 | 100 | 74.60  | 350  | 261.1  |
| 10 | 7.460 | 36 | 26.856 | 110 | 82.06  | 400  | 298.4  |
| 11 | 8.206 | 38 | 28.348 | 120 | 89.52  | 450  | 335.7  |
| 12 | 8.952 | 40 | 29.840 | 130 | 96.98  | 500  | 373.0  |
| 13 | 9.698 | 42 | 21.332 | 140 | 104.44 | 600  | 447.6  |
| 14 | 10.444 | 44 | 32.824 | 150 | 111.90 | 700  | 522.2  |
| 15 | 11.190 | 46 | 34.316 | 160 | 119.36 | 800  | 596.8  |
| 16 | 11.936 | 48 | 35.808 | 170 | 126.82 | 900  | 671.4  |
| 17 | 12.682 | 50 | 37.300 | 180 | 134.28 | 1000 | 746.0  |
| 18 | 13.428 | 55 | 41.03  |     |        |      |        |

*Courtesy Allis-Chalmers Co.*

Circuits serving outside outlets must be protected by ground-fault interrupter equipment which protects against accidental shock.

Interior convenience outlets should be installed 18 inches above the floor in bedrooms, living and dining rooms. In the dining room place one wall outlet above the table height to connect portable cooking devices from the table. Install work area and kitchen outlets 8 inches above counter or work table levels. Laundries, kitchens and home work shops should have lighting directly over the work areas.

Individual power circuits of 240 volts are necessary for major appliances including wall ovens, clothes dryers, electric ranges, space heaters, and large air conditioning units.

### Table 4. Dimensions, Weight, and

| Gauge No. A.W.G. | Dia. in. | Area Circular Mils (d²) (1 Mil = .001 in.) | Lbs. per 1000 Feet Bare Wire | Length (Feet per lb.) | Resistance at 77° F. (Ohms per 1000 ft.) |
|---|---|---|---|---|---|
| **Stranded** | 1.151 | 1000000. | 3090. | .3235 | .0108 |
| | 1.029 | 800000. | 2470. | .4024 | .0135 |
| | .963 | 700000. | 2160. | .4628 | .0154 |
| | .891 | 600000. | 1850. | .5400 | .0180 |
| | .814 | 500000. | 1540. | .6488 | .0216 |
| | .726 | 400000. | 1240. | .8060 | .0270 |
| | .574 | 250000. | 772. | 1.30 | .0431 |
| 0000 | .4600 | 211600. | 640.5 | 1.55 | .0500 |
| 000 | .4096 | 167800. | 507.9 | 1.97 | .0630 |
| 00 | .3648 | 133100. | 402.8 | 2.48 | .0795 |
| 0 | .3248 | 105500. | 319.5 | 3.13 | .1002 |
| 1 | .2893 | 83690. | 253.3 | 3.95 | .1264 |
| 2 | .2576 | 66370. | 200.9 | 4.98 | .1593 |
| **Solid** 3 | .2294 | 52640. | 159.3 | 6.28 | .2009 |
| 4 | .2043 | 41740. | 126.4 | 7.91 | .2533 |
| 6 | .1620 | 26250. | 79.46 | 12.58 | .4028 |
| 8 | .1284 | 16510. | 49.98 | 20.01 | .6405 |
| 10 | .1018 | 10380. | 31.43 | 31.82 | 1.018 |
| 12 | .0808 | 6530. | 19.77 | 50.59 | 1.619 |

Circuits of 20 amperes at 120 volts are minimum requirements in the kitchen and workshop to provide negligible interruption for appliance use.

## Indoor Lighting

Providing enough lighting on the interior of buildings, particularly homes, is not always an easy task. Enough light means providing sufficient light for a person to see quickly and easily. Supplying a good light source is essential where you read, study, at the bathroom mirror, and in locations where good seeing is necessary.

Lighting should be comfortable. The right kind of quality lighting is an important as providing enough light. Avoid installing glare type lighting.

## Resistance of Pure Copper Wire

| | Gauge No. A.W.G. | Dia. in. | Area Circular Mils ($d^2$) (1 Mil = .001 in.) | Lbs. per 1000 Feet Bare Wire | Length (Feet per lb.) | Resistance at 77° F. (Ohms per 1000 ft.) |
|---|---|---|---|---|---|---|
| | 14 | .0640 | 4107. | 12.43 | 80.44 | 2.575 |
| | 16 | .0508 | 2583. | 7.82 | 127.90 | 4.094 |
| | 18 | .0403 | 1624. | 4.92 | 203.40 | 6.510 |
| | 20 | .0319 | 1022. | 3.09 | 323.4 | 10.35 |
| | 22 | .0254 | 642. | 1.95 | 514.2 | 16.46 |
| | 24 | .0201 | 404. | 1.22 | 817.7 | 26.17 |
| | 26 | .0159 | 254. | .77 | 1300. | 41.62 |
| | 28 | .0126 | 159.8 | .48 | 2067. | 66.17 |
| | 30 | .0100 | 100.5 | .30 | 3287. | 105.2 |
| | 32 | .0080 | 63.2 | .19 | 5227. | 167.3 |
| Solid | 34 | .0063 | 39.7 | .12 | 8310. | 266.0 |
| | 36 | .0050 | 25.0 | .076 | 13210. | 423.0 |
| | 38 | .0040 | 15.7 | .047 | 21010. | 672.6 |
| | 40 | .0031 | 9.89 | .030 | 33410. | 1069.0 |
| | 42 | .0025 | 6.22 | .019 | 52800. | 1701. |
| | 44 | .0020 | 3.91 | .012 | 82500. | 2703. |
| | 46 | .0016 | 2.46 | .008 | 128800. | 4299. |
| | 48 | .0012 | 1.55 | .004 | 229600. | 6836. |
| | 50 | .0010 | 0.97 | .003 | 330000. | 10870. |

Fluorescent lighting is recommended by lighting experts where feasible. Size of this type is determined by the fixture or equipment in which it is used. Two types of tubes are generally used: deluxe warm white (WWX) and deluxe cool white (CWX). Warm white improves warm light colors and flatters the complexion. Cool white bulbs enhance the deeper color scheme tones including the blue and/or green shades if these are used in decorating.

Light bulbs may carry the stamped-on word LUMENS which means the LIGHT OUTPUT—you initially get out of a bulb. Fluorescent tubes are more efficient than incandescent bulbs: 40-watt fluorescent = 2080 initial lumens, 40-watt incandescent = 450 initial lumens. It must be remembered however, that some incandescent bulbs are more efficient than others.

Examples of methods used to calculate recommended indoor lighting, electrical circuits, feeders, and main building electrical entrances are found in the National Electrical Code, a copy of which may be purchased from the National Fire Protection Association, 470 Atlantic Avenue, Boston, MA 02210.

Other helpful guide information may be secured from the National Electrical Manufacturers Association, 155 E. 44th St., New York, N.Y. 10017.

## ELECTRIC SERVICE

The power-line wiring and the building service panel are the major factors in assuring efficient operation of building lighting, fixtures, appliances, and machinery. Problems will occur if the service wires are too small and if the service panel is undersize.

150- and 200-amp service require three 1/0 or 3/0 wire with RHW insulation. This type of service is usually installed for small shops, ranches, farms, and larger homes. The 200-amp service is recommended for large ranches and farms, and for small shops where considerable machinery is installed.

In electrical wiring, the National Electrical and local codes must be complied with. Occasionally the local power company has regulations over and above those of local and national codes. These regulations also must be complied with. Materials used should be UL approved for interior work and Power Company approved for exterior work.

## GROUNDING

The white or neutral wire of all alternating current electrical systems must be grounded. A No. 6 or No. 4 copper ground wire, free of mechanical damage, is normally used. No. 8 wire must be armored, or if bare, should be conduit enclosed. Metal raceways, conductor enclosures, and electrical equipment also must be grounded. Grounding prevents shocks and reduces the effects of lightning and high-voltage surges.

In major buildings and on farms, the ground wire is not processed through the entrance switch but is taken directly from a neutral supply or overhead wire. The ground is tied to an underground metallic water-pipe system with a ground clamp. If water piping is not available for ground wire attachment, an 8-foot rod at least 1/2-inch in diameter should be placed at least 2 feet from the building and driven into the ground at least 12 inches below the surface before the ground wire is attached and clamped.

## TRANSFORMERS

*(Information Courtesy Allis-Chalmers Co.)*

The transformer is an important electrical device by which the energy in alternating current circuits can be received at one voltage and delivered at a lower or a higher voltage. A transformer has no moving parts. The principal components are an *iron core* which provides a path of low reluctance for the magnetic flux, the *primary winding* which receives energy from the supply source, and the *secondary winding* which receives energy by induction from the primary source and delivers it to the secondary circuit.

Pole-mounted transformers are normally filled with a special transformer oil that completely surrounds the winding and the core.

### Overload Protection

Transformers are protected against overloading in several ways. One method is to mount a simple fuse on the pole above the transformer. Since the fuse rating is coordinated with the transformer capacity, an overload will blow the fuse and remove the transformer from the line. This fuse is generally sized to protect the transformer in case of short circuits and harmful overloads, although many utilities use a minimum size fuse regardless of the transformer size.

## Operation

Transformers should be installed at least 18 inches from other transformers, from adjacent walls, partitions, etc., to permit free air circulation. Voltage should be applied only to transformers filled with oil to the correct level. Transformers should be protected from excessive overloads and overvoltage surges with approved protective devices. The total load on a transformer should not exceed the nameplate rating for a continuous load in a standard ambient temperature.

If the unit has a tap changer, operate it by turning the handle until the pointer is above the letter indicating the desired tap. The tap changer snaps into each tap position with an audible click.

## Precautions

For safety to the transformer and to those working on the equipment, the tank or case must have a solid, permanent ground. Where bases are not welded to the tank, the tank itself should be grounded. However, if used on a system where ungrounded tank practice is common, a tank discharge gap must be placed between the tank and a solid ground. Provisions for grounding are always supplied on the equipment. Grounding of neutral bushings depends on the system to which the transformer is connected.

When any tests are being made, have only a qualified test man on the transformer. Treat all wires and testing equipment as if they were energized and capable of causing a severe shock.

## Safety Precautions

All safety rules of your company should be followed. Do not take a chance. Work safely.

Make certain the transformer is de-energized and cannot be re-energized while you are working on it.

Use only fire extinguishers approved for electrical equipment fires. Two such types generally available are $CO_2$ and dry chemical.

Your own life and the lives of others depend on the care you use.

CHAPTER 9

# Air Conditioning and Refrigeration

Refrigeration, generally, is the process of removing heat from a specific location. Heat is removed from the air in a room for the purpose of establishing human comfort, from food to preserve the quality and flavor, and from general locations to accomplish or establish a desired effect.

Refrigeration processes mechanically transfers unwanted heat to unobjectionable areas. This is accomplished by the liquid substance in the refrigeration machine, called the *refrigerant*. In effect, a refrigerant is any liquid which will evaporate and boil at a low temperature. The refrigerant absorbs the heat during the act of evaporation or boiling.

When operating, a refrigeration unit allows the refrigerant to boil in tubes which are in direct or indirect contact with the air or other medium to be cooled. Proper engineering design and controls determine the operating temperatures of the refrigeration machine or unit.

## THE REFRIGERATION CYCLE

The process of refrigeration may be more readily understood by following a typical refrigeration cycle while referring to the diagram shown in Fig. 1.

The liquid refrigerant begins its travels from the storage container called the *receiver*. The temperature of this liquid refrigerant is usually slightly above room temperature. In view of the low boiling point of commercial refrigerants, the pressure generated by the liquid in the receiver is sufficient to cause the liquid to travel through the *liquid line* to the *metering device.* The metering device may be a capillary tube or a type of expansion valve. It has the effect of reducing the pressure of the refrigerant as the liquid enters the *evaporator coil.*

Immediately after passing through the metering device, the refrigerant still exists largely in the liquid state. However, as a

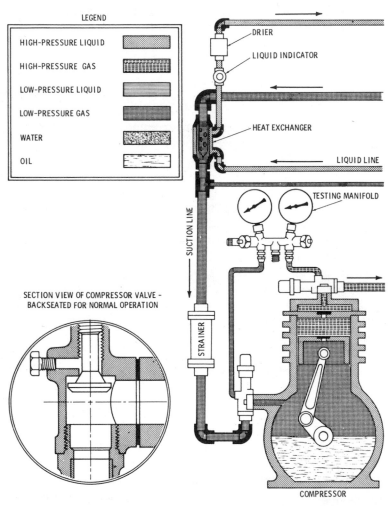

**Fig. 1. Diagram of**

result of the pressure reduction, it is now at a considerably lower temperature than it was before entering the metering device. Heat from the medium being cooled passes through the walls of the evaporator and is absorbed by the refrigerant. More and more of the low-temperature liquid boils and changes into a gas, so that by the time the refrigerant reaches the evaporator outlet, it has all been vaporized.

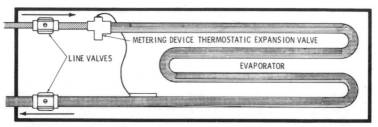

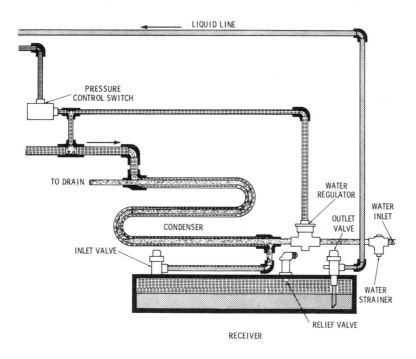

*Courtesy Mueller Brass Co.*

*a refrigeration cycle.*

The evaporator need not be in the form of a serpentine coil, as illustrated in Fig. 1. In household refrigerators and milk coolers, for example, the evaporator is in the form of a plate, through which the refrigerant flows. Passages are blown or shaped into the plate.

From the evaporator outlet, the refrigerant vapor is drawn through the *suction line* into the *compressor* by the action of the much lower pressure created on the low side of the compressor. In the compressor (also referred to as the *pump*), the low-pressure, low-temperature refrigerant vapor is compressed into high-pressure, high-temperature vapor. This high-temperature vapor passes into the *condenser* where it is cooled by the action of water or air surrounding the coil walls. This cooling action causes the high-pressure gas to condense or liquify. The liquid refrigerant flows into the receiver where it is ready once more to repeat the cycle just described.

It is important to remember that the compressor does not pump or push the liquid refrigerant. The liquid refrigerant flows through the liquid line to the metering device by virtue of its own natural pressure. The low-temperature, low-pressure vapor issuing from the evaporator is pulled back by the compressor.

In the condenser, the heat which the refrigerant picked up from the medium being cooled is transferred to the atmosphere in the case of an air-cooled condenser, or transferred to water which may be either drained into the sewer or reclaimed by the use of a cooling tower.

Basically, then, the refrigeration cycle consists of alternately vaporizing and condensing the refrigerant. In so doing, it is possible to absorb heat in the refrigerant at the evaporator and reject the heat from the refrigerant at the condenser. The refrigerant undergoes no chemical change. Unless a leak develops in the system, the refrigerant is reusable indefinitely.

The condenser, receiver, and liquid line of a system are jointly referred to as the *high side* of the unit. The evaporator and suction line are called the *low side* because the pressures are much lower than in the high side. The physical assembly which contains the compressor, condenser, and receiver is called the *condensing unit*.

## AIR CONDITIONING

Air conditioning is the process of treating air so as to control its temperature, humidity, cleanliness, and distribution to meet the requirements of a given space or area. If these requirements are set for human occupants, the process is termed *comfort conditioning*. If the requirements are for an industrial procedure, air conditioning is called *process conditioning*. Comfort conditioning includes installations in homes, theaters, restaurants, offices, and transportation facilities. Textile mills, electronic-component assembly areas, candy plants, etc., make use of industrial or process conditioning.

The term "air conditioning" is commonly applied only to summer cooling. However, it more properly includes both summer and winter processes. In winter, humidifiers may be added to the heating equipment to maintain a comfortable range of moisture in the air. In summer, the opposite is usually true, and it is necessary to dehumidify the air to bring the relative humidity within a comfortable range. In both summer and winter, ventilation must be considered to maintain an adequate supply of oxygen and to reduce the concentration of disagreeable odors, including those due to smoking. Studies have indicated that personnel efficiency increases in air-conditioned offices. Absenteeism drops and other intangible benefits result when comfort conditioning is installed.

Until recently, almost all large air-conditioning installations made use of water-cooled condensers. To save water, water towers or evaporative condensers are used in such installations. These devices recirculate most of the water, cooling it by exposing it to the atmosphere. Water problems have increased in most sections of the country, however, because of the low supply or because of poor quality. This factor is responsible for a rapidly increasing use of air-cooled condensers.

A relative newcomer to air conditioning is the *heat pump*. This is basically an air conditioner redesigned and recirculated so that, by the flip of a heating-cooling control switch, a cooling unit becomes a heating unit. What actually happens is that a reversing valve changes the flow of refrigerant (except through the compressor) to make the evaporator the condenser and the condenser the evaporator during the heating cycle. While water-source heat pumps (those which obtain and reject heat from and to a well or lake) are the most efficient and economical to operate, water problems have forced more general acceptance of the air-to-air heat pumps which extract heat even from winter air to heat the inside of the facility or building. Ground-source heat pumps, which obtain their heat through a refrigerant coil buried in the earth, have been found impractical in all but unique or unusual applications.

The Mueller Brass Company of Port Huron, Michigan, has played a significant role in the development of mechanical refrigeration and have developed dehydrated copper tubing, fittings, valves, and accessories designed specifically for the extensively used refrigerants 12 and 22. These fittings are being used in all phases of air conditioning and refrigeration work with great success. Refrigerants will leak through ordinary copper fittings and tubing capable of retaining high-pressure water. For this reason, all copper lines and fittings in refrigerating systems must be absolutely seepage-proof and free of any restrictions which might impede the flow of refrigerant.

## REFRIGERANT LINE SIZES

Correct line sizes are essential to obtain maximum efficiency from any refrigeration equipment. In supermarkets, for example, the long lines running under the floor from the display cases to the machine room to the rear of the store must be fully engineered. Otherwise, problems of oil return, slugging, or erratic refrigeration are quite likely.

Where they are available, the manufacturer's recommendations must be followed relative to step-sizing risers, traps, and like items. For average conditions, Table 1 may be used to determine the size of the liquid and suction lines. For runs over 100 ft. but under 150 ft., the next larger size of tubing should be used; for runs over 150 ft., use tubing that is two sizes larger than shown in Table 1.

## Table 1. Sizes of Refrigerant Lines

| Btu. Per Hour | REFRIGERANT 12 | | | REFRIGERANT 22 | | | REFRIGERANT 40 | | |
|---|---|---|---|---|---|---|---|---|---|
| | Liquid Line | Suction Line 5° F | Suction Line 40° F | Liquid Line | Suction Line 5° F | Suction Line 40° F | Liquid Line | Suction Line 5° F | Suction Line 40° F |
| 3,000 | 1/4 | 1/2 | 1/2 | 1/4 | 1/2 | 1/2 | 1/4 | 1/2 | 1/2 |
| 6,000 | 3/8 | 5/8 | 5/8 | 3/8 | 5/8 | 5/8 | 1/4 | 1/2 | 1/2 |
| 9,000 | 3/8 | 7/8 | 5/8 | 3/8 | 7/8 | 5/8 | 3/8 | 5/8 | 5/8 |
| 12,000 | 3/8 | 1 1/8 | 7/8 | 3/8 | 7/8 | 7/8 | 3/8 | 7/8 | 7/8 |
| 15,000 | 3/8 | 1 1/8 | 7/8 | 3/8 | 1 1/8 | 7/8 | 3/8 | 7/8 | 7/8 |
| 18,000 | 3/8 | 1 1/8 | 7/8 | 3/8 | 1 1/8 | 7/8 | 3/8 | 1 1/8 | 7/8 |
| 21,000 | 1/2 | 1 1/8 | 1 1/8 | 1/2 | 1 1/8 | 1 1/8 | 3/8 | 1 1/8 | 7/8 |
| 24,000 | 1/2 | 1 3/8 | 1 1/8 | 1/2 | 1 1/8 | 1 1/8 | 1/2 | 1 1/8 | 7/8 |
| 30,000 | 5/8 | 1 3/8 | 1 1/8 | 1/2 | 1 3/8 | 1 1/8 | 1/2 | 1 1/8 | 1 1/8 |
| 36,000 | 5/8 | 1 3/8 | 1 1/8 | 5/8 | 1 3/8 | 1 1/8 | 1/2 | 1 3/8 | 1 1/8 |
| 42,000 | 5/8 | 1 5/8 | 1 3/8 | 5/8 | 1 3/8 | 1 3/8 | 1/2 | 1 3/8 | 1 1/8 |
| 48,000 | 5/8 | 1 5/8 | 1 3/8 | 5/8 | 1 5/8 | 1 3/8 | 1/2 | 1 3/8 | 1 1/8 |
| 54,000 | 5/8 | 1 5/8 | 1 3/8 | 5/8 | 1 5/8 | 1 3/8 | 5/8 | 1 3/8 | 1 1/8 |
| 60,000 | 7/8 | 1 5/8 | 1 3/8 | 5/8 | 1 5/8 | 1 3/8 | 5/8 | 1 5/8 | 1 3/8 |
| 72,000 | 7/8 | 2 1/8 | 1 5/8 | 7/8 | 1 5/8 | 1 3/8 | 5/8 | 1 5/8 | 1 3/8 |
| 96,000 | 7/8 | 2 1/8 | 1 5/8 | 7/8 | 2 1/8 | 1 5/8 | 5/8 | 2 1/8 | 1 5/8 |
| 108,000 | 7/8 | 2 5/8 | 2 1/8 | 7/8 | 2 1/8 | 1 5/8 | 7/8 | 2 1/8 | 1 5/8 |
| 120,000 | 7/8 | 2 5/8 | 2 1/8 | 7/8 | 2 1/8 | 1 5/8 | 7/8 | 2 1/8 | 1 5/8 |
| 150,000 | 1 1/8 | 2 5/8 | 2 1/8 | 7/8 | 2 1/8 | 2 1/8 | 7/8 | 2 1/8 | 2 1/8 |
| 180,000 | 1 1/8 | 2 5/8 | 2 1/8 | 1 1/8 | 2 5/8 | 2 1/8 | 7/8 | 2 5/8 | 2 1/8 |
| 210,000 | 1 1/8 | 3 1/8 | 2 1/8 | 1 1/8 | 2 5/8 | 2 1/8 | 7/8 | 2 5/8 | 2 1/8 |
| 240,000 | 1 3/8 | 3 1/8 | 2 5/8 | 1 3/8 | 2 5/8 | 2 1/8 | 7/8 | 2 5/8 | 2 1/8 |
| 300,000 | 1 3/8 | 3 1/8 | 2 5/8 | 1 3/8 | 3 1/8 | 2 5/8 | 1 1/8 | 2 5/8 | 2 1/8 |
| 360,000 | 1 3/8 | 3 5/8 | 2 5/8 | 1 3/8 | 3 1/8 | 2 5/8 | 1 1/8 | 3 1/8 | 2 5/8 |
| 420,000 | 1 5/8 | 3 5/8 | 3 1/8 | 1 3/8 | 3 1/8 | 2 5/8 | 1 1/8 | 3 1/8 | 2 5/8 |
| 480,000 | 1 5/8 | 4 1/8 | 3 1/8 | 1 5/8 | 3 5/8 | 3 1/8 | 1 1/8 | 3 1/8 | 2 5/8 |
| 540,000 | 1 5/8 | 4 1/8 | 3 1/8 | 1 5/8 | 4 1/8 | 3 1/8 | 1 3/8 | 3 5/8 | 3 1/8 |
| 600,000 | 1 5/8 | 4 1/8 | 3 1/8 | 1 5/8 | 3 5/8 | 3 1/8 | 1 3/8 | 3 5/8 | 3 1/8 |

## AIR-CONDITIONING UNITS

### Window Units

Many small, self-contained, air-conditioning units are installed in homes, offices, and other locations today. They are complete with compressor, motor and drive, evaporator, air-cooled condenser, fans, filters, and controls, all enclosed in a cabinet and factory assembled. Window units are normally not designed for automatic operation and a switch is therefore provided for manual operation. Approximate areas that are effectively cooled by this type of unit are listed in Table 2.

### Table 2. Cooling Capacities of Typical Self-Contained Air-Conditioning Units.

|  | Average Btu-per-hour Capacity | Floor Area (sq. ft.) |
|---|---|---|
| ½-HP window mounted | 5,500 | 140 |
| ¾-HP floor mounted | 7,500 | 190 |
| ¾-HP window mounted | 9,000 | 225 |

Window units are mounted directly in a window opening, while the small floor-model units are placed on the floor in front of a window with a metal duct extension to the window. Complete instructions and materials are included with window-mounted units for installing them in most sizes and types of windows.

### Floor-Mounted Units

Floor-mounted air-conditioning units normally are of a capacity ranging from 3 to 7-1/2 hp and are complete with evaporator. blower, filters, compressor, water-cooled condenser, and accessories, all factory assembled in an upright prepainted cabinet. These units normally are installed in the area to be airconditioned and require only a small amount of duct work.

### Maintenance

If the units are operated daily, the filters should be cleaned and/or replaced at least every other week during the use season. At least once each year the unit should be completely inspected, including the checking of switches, louvers, coils, fan assembly (including the fan drive belt and fan motor), general control system, and grilles and dampers if they are existing. The unit should be tested for refrigerant leaks around the receiver, service valves, valve caps, and connections.

## INSPECTION AND PREVENTIVE MAINTENANCE

The following procedures and inspections should be performed on any type of air-conditioning system at least every 3 months.

*Electric Power Supply*—Inspect the electric power-supply source switch, electrical connections, and limit switches for possible fraying or loose connections.

*Louvers*—Clean dirt, dust, and lint from all louvers and bird screens.

*Cooling Coil*—Inspect the cooling-coil supports for rust spots, and paint if necessary. Clean the coils and fins.

*Heating Coils*—Inspect steam-trap operation, clean steam lines and drip strainers, inspect hand and automatic valves for proper operation and leaks, and inspect heat coils for leaks.

*Fan Assembly*—Inspect fan operation for excessive vibration, bent blades, and proper shaft-bearing lubrication. Be careful not to overlubricate the shafts. Wipe all dirt from bearings, fan wheel, and shaft.

*Fan Drive*—Inspect the fan drive for belt tension, condition of belts, and pulley alignment. Wipe all dirt and oil from the belts and pulleys.

*Fan Motor*—Clean the dirt from the motor, inspect for lubrication requirement, and observe for any unusual appearances and sound.

*Fan Switches*—Inspect contact points and operation.

*Casing*—Inspect insulation for loose or missing sections, thoroughly clean all sections and inspect for rust, applying touch-up paint if required.

*Control System*—In both large and small systems, inspect the operation of thermostats, humidistats, automatic valves, damper operators, limit switches, and relays, as may be applicable. On pneumatic control systems, inspect the air compressor, drive, tank, motor, pressure regulator, and operating pressures.

*Duct Work*—Inspect the settings of grille louvers and dampers for proper positioning. Lubricate damper bearings and inspect automatic dampers for freedom of operation. If duct work is insulated, loose and missing insulation should be refastened and/or replaced as conditions warrant.

The following inspection-maintenance procedures should be performed in conjunction with refrigerated mechanical cooling systems.

*Operation*—Inspect equipment for any abnormal sounds or vibrations while the equipment is in operation. Note and correct any apparent mechanical deficiencies.

*Expansion Valves*—Inspect the operation of all expansion valves, installing a service dryer in the liquid line if moisture is suspected.

*Condenser*—Check water-cooled unit connections for water leaks and rust spots. If the unit is air-cooled, clean the coil, fins, and fan blades.

*Refrigerant Charge*—Inspect for proper refrigerant charge if there is a possibility that it may be low, making sure that all refrigerant lines have been tested for leaks. When inspecting lines for leaks, pay particular attention to the connections and the equipment containing the refrigerant.

*Compressor Motor, Drive, and Body*—Check the motor bearings for lubrication requirement and wipe dirt from housing. Inspect the drive for belt tension and condition of belts, and check pulley alignment. Clean lint, grease, and dirt from flywheel, belt, and pulleys. On the compressor body, inspect the oil level in the compressor crankcase, and inspect the seals and gasket for possible failure or leaks.

Inspection and general maintenance for ventilation and summer-use exhaust systems are similar in most respects to those used in conjunction with air-conditioning equipment, except that no cooling and/or refrigeration equipment exists in most instances.

For evaporative-type cooling systems, the general maintenance and inspection procedures listed for air conditioning should be followed. In addition, particular attention should be given to the following items.

*Water Supply and Drains*—Inspect the supply and drain lines and fittings for possible leaks, making any repairs that are required. Inspect and repair (if required) the float-valve setting to assure proper water overflow. Clean the drain-line connection and inspect the float-valve operation.

*Spray System and Pads*—Inspect and clean spray nozzles and troughs, and inspect piping connection for possible leaks. Inspect pads for possible damage; clean off excess scale or dirt, replacing the pads only if necessary.

Every attempt should be made to conduct maintenance inspections and services so as to interfere as little as possible with user activities. However, inspections must be made periodically to insure continuous operation and avoid major equipment breakdowns. The type of equipment, manufacturer's recommendations, and equipment use will determine the frequency of maintenance and/or preventive maintenance inspections. Inspection records should be maintained on all equipment.

## SAFETY PRECAUTIONS

It is necessary to avoid bodily contact with liquid refrigerant and to avoid inhaling refrigerant gas. Care must be used so that refrigerants do not contact the eyes. In case of refrigerant leaks, immediately ventilate the area.

Before repairing any machinery, be sure that all power is off, and do not adjust, clean, or lubricate parts which are in motion unless such parts have proper guards or are fully enclosed.

## HOME COOLING AND VENTILATING FACTORS

1. Central cooling equipment may be installed in an attic or crawl space freeing living space areas.
2. Light colored roofing materials reflect much more of the sun's heat than darker materials. This keeps the home cooler.
3. Shade the house against the direct rays of the sun with trees, awnings, and other natural or artificial shading. This also helps to keep the home cooler.
4. Large windows facing north and south are cooler than those facing the east and west.
5. For a low cost cooling aid, ventilation fans can be installed in the attic.

6. Install air duct distribution systems in straight runs. Avoid sharp turns as these create air flow resistance reducing air flow distribution.

7. Concrete or blacktop areas adjacent to a house reflect or radiate heat into the house making it more difficult to keep the house cool.

8. Leaving storm windows on reduces the air-conditioning load, and helps keep out infiltrating air and some outside noise.

9. Room air conditioners in windows cost less than central air systems but are noisier and provide less uniform temperatures. However, with window-installed units the house or building can be air conditioned at desired locations at less cost and the amount of cooling in various rooms with these units can be varied.

10. If mechanical air conditioning is not planned, consider cross ventilation. Shrubs and trees should be arranged to provide shade without limiting breezes. Use deciduous trees (those which lose their leaves) letting in winter sunshine.

11. Awnings, louvered bar screens and certain window shades help keep out the summer sun easing the excessive indoor cooling load.

# CHAPTER 10

# Insect and Rodent Control

Insects and rodents, if not effectively controlled, have the capability of destroying vast quantities of edible products, plant and tree life, and endangering human and animal health and life. Not only is it necessary to destroy the many types of unwanted insects and rodents, but preventive measures are required to eliminate the possibility of infestation.

A major factor in insect and rodent preventive control is sanitation. All items of waste which attract insects and rodents must be adequately covered and disposed of in a manner recommended by responsible authorities in your particular area. High standards of sanitation must be continuously observed.

There are many types of chemicals and products used in pest control operations and also many application procedures.

The status of many insecticides and herbicides change. It is suggested that when a problem exists, your local supply dealer be consulted as he keeps informed of the latest restrictions and recommendations. You will also find the State Extension Entomologist or Extension Service cooperative.

## DISPENSERS

The most common type of dispenser for pest control material is the hand-operated aerosol "bomb" which contains a solution of insecticide and liquified gas. The gas provides the pressure to dispense the insecticide in a very fine spray.

*Fig. 1. A hand-operated sprayer.*

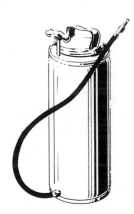

*Fig. 2. A compressed-air sprayer.*

*Fig. 3. A plunger-type hand duster.*

For general spraying purposes, the most common types of equipment used are the hand sprayer (Fig. 1), compressed-air sprayer (Fig. 2), and the plunger-type hand duster (Fig. 3). All are readily available at most hardware and supply stores.

## INSECTICIDES

There are many types of insecticides in use including the following:

### DDT (dichloro-diphenyl-trichloroethan)

*DDT* is a chlorinated hydrocarbon insecticide, and is a very useful chemical for insect control. It is highly toxic to insects, both by ingestion and by contact, and has a great degree of persistence.

### Allethrin

*Allethrin* is a synthetic complex pyrethrin-like compound, and is used extensively in some locations for control of adult flies and mosquitoes.

### Lead Arsenate

*Lead arsenate* is used primarily as a stomach poison to control chewing insects, such as grubs, weevils, etc., that eat the foliage, bark, or roots of plants and shrubs.

### Lindane

*Lindane* is very effective as a stomach poison and as a residual insecticide. It may be used for some residual application indoors, as a larvicide where resistant house flies are encountered, and as a space spray.

### Chlordane

*Chlordane* is toxic to most insects and acts to a considerable degree as a fumigant. It controls practically every species of problem insects found in the United States. It is widely used by professional pest control operators for home, commercial, industrial, and institutional insect control. It is economical, easy to apply, and gives good results.

Chlordane insecticide is being replaced by a new insecticide for insect control—GOLD CREST *FICAM-W*. Chlordane is to be removed from normal sales sources after shelf supplies in the various stores and supply houses are exhausted.

## INSECT CONTROL

The following is a general listing of the methods used to control most common insects.*

### Ants

There are many species of ants, but the most common pest of lawn, garden, and household is the little black ant illustrated in

*Fig. 4. The common black ant.*

Fig. 4. The length of this ant is from 1/12″ to 1/10,″ with a slender body, jet black, and shiny. Ants usually nest in the soil or in the masonry or woodwork of buildings. Some nests are hard to find as they are located under sidewalks or driveways, under boards or stones, and at times in decaying logs or tree trunks.

Ants in the lawn are an irritating, unsightly nuisance, and many have painful bites. They injure flowers and vegetables by direct feeding or by carrying aphids (plant lice) from plant to plant, and destroy vegetation around their nests. In the home, ants may cause expensive losses by spoiling food.

In buildings, spray or dust the points at which the ants enter with chlordane. In gardens, treat the entire infested area with *Dursban* spray or dust. *Do not apply to edible portions of fruits or vegetables after they have formed.* In lawns, treat the hills with spray or dust. If the ants are numerous, treat the entire lawn, and water in.

### Army Worms

Army worms (Fig. 5) are approximately 1-1/2″ long, and are greenish-brown in color with brown, orange, and white longitudinal stripes. The head is honeycombed with dark lines, and

---

*Information and illustrations courtesy Velsicol Chemical Corporation.

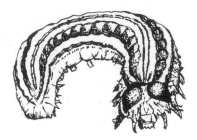

*Fig. 5. Army worm.*

each fore leg has a dark band on the outer side and a dark tip on the inner side.

The damage caused by army worms consists primarily of the insect eating the plant foliage.

To control, treat the soil with lead arsenate or *sevin* before planting, and the soil surface around the base of young seedlings after planting.

## Asiatic Garden-Beetle Larvae

The larva of the Asiatic Garden Beetle (Fig. 6) is similar in appearance to white grubs, but with a more slender body having a Y-shaped opening at one end behind a transverse row of setae (short bristles). The larvae are from 1/2″ to over 1″ in length,

*Fig. 6. Larva of the Asiatic garden beetle.*

and they feed on roots. The adult beetles feed on leaves and buds, and on the flowers of vegetable, fruit, berry, and ornamental plants. To control this pest, treat the soil in early spring (before planting) with chlordane.

## Box-Elder Bugs

Box-elder bugs (Fig. 7) have a narrow body, a flat back, and are about 1/2″ long. Their color is brownish black with three longitudinal red stripes on the thorax and red veins on the wings.

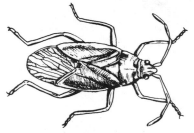

*Fig. 7. Box-elder bug.*

These bugs feed on flowers, fruits, foliage, and twigs of box-elder, ash, and other trees. They also invade the home and other buildings.

To control the box-elder bug, dust or spray lead arsenate or *sevin* around door sills, window frames, and other points at which they may enter the building. Also treat those areas where the insects crawl or congregate, such as porches, side walls of houses, box-elder trees, and similar locations.

## Carpet Beetles

Carpet beetles (Fig. 8) are elliptically shaped, 5/32″ long, and 75% wider than they are long. A band of brick-red scales runs down the center of their back with red and white scales forming

*Fig. 8. Carpet beetle.*

irregular bands about the body. Some species encountered may have a dull black body with brown legs and antennae.

The adult carpet beetle is harmless, but the larvae will eat holes in clothing, rugs, furniture padding, or any material containing wool, feathers, hair, or fur.

To control, treat the breeding places, such as the underside of floor coverings, cracks in floors and baseboards, folds in upholstery, etc., with lead arsenate.

### Chiggers

Chiggers (Fig. 9) are bright orange-yellow, oval, blind, six-legged mites less than 1/150″ in diameter. They are almost invisible to the naked eye and are usually found on vegetation. These pests insert their mouth parts into human pores or hair

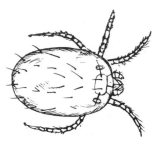

**Fig. 9.  Chigger.**

folicles and release poison which causes irritation and severe itching. Welts and severe itching may last as long as a week.

All lawn areas known to be infested should be treated with *sevin* or other approved insecticide.

### Centipedes

Centipedes (Fig. 10) are from 2″ to 3″ long when full grown, and are a grayish-tan color. They have many long legs extending

**Fig. 10.  Centipede.**

around their body, and have a long antenna. Centipedes are considered harmless by many people, but they can inflict a painful "bite" with a pair of poison claws that are located just behind their head.

To eliminate these insects, all areas around water pipes or other damp parts of the house or building should be treated with *sevin*.

## Cloths Moths

Cloths moths (Fig. 11) are small, buff-colored winged insects with a wing span of 1/2". Their eggs are white and about the

*Fig. 11. Clothes moth.*

size of a pinhead. Under warm conditions, the eggs will hatch in from 4 to 9 days; in colder conditions, sometimes not for 15 days. The moths themselves do no damage, but their larvae feed on wool, furs, feathers, hair, and fabrics made from these materials. The adult (winged) specimens should be killed as each moth may lay 100 or more eggs during its life.

Treat all fabrics, fabric storage areas, and warm- and cold-air passages with moth repellent such as napthalene. Dry clean or properly air in sunlight and brush articles before they are stored or used. Moth balls are an excellent control for moths.

## Cockroaches

The most common of the several species of cockroaches (Fig. 12) are tan, 1/2" to 1" long, and have two dark stripes on the upper side. They do not often show themselves during daylight and may be present in considerable numbers before their existence in a building is realized. Cockroaches feed on various food products and paper, and are said to carry diseases.

*Fig. 12. Cockroach.*

Thoroughly spray or dust all infested cracks and other hiding places, as well as adjacent exposed surfaces where the roaches will crawl when they come out of hiding. Repeat the process as often as required.

## Crickets

The size, color, and wing length of crickets (Fig. 13) vary. Adult crickets usually have a body that is 3/5″ to 1″ long with an antennae half as long as the body. All crickets have heavy hind legs. Crickets can be highly destructive, as they will eat almost anything, including leather, fabrics, bookbindings, plant foliage, flowers, and tender-growth plants.

*Fig. 13. Cricket.*

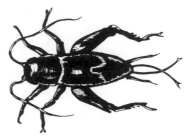

To control crickets, treat all lawn surfaces and building foundations. Spray or dust baseboards, floors, closets, storage areas, and other hiding places indoors. Repeat the application as often as necessary.

## Grasshoppers

There are wide variations in the color and size of grasshoppers (Fig. 14). Most, however, are from 1″ to 2″ long when fully grown. This insect is easily recognized by its prominent rear jumping legs. The only damage caused by grasshoppers is to plant foliage.

*Fig. 14. Grasshopper.*

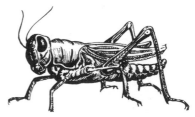

To control grasshoppers, treat growing plants before blooms appear, and treat grassy or weedy areas around gardens.

## Earwigs

Earwigs (Fig. 15) are brown or blackish in color with a pincer-like appendage at the rear. They breed in trash, compost, lawn clippings, and similar trash areas. Earwigs inflict damage on vegetables, fruits, and flowering plants, and will often invade buildings.

To control this pest, treat lawn surfaces, trash areas, around fences and foundations, and around compost heaps and flower beds. Do not sprinkle or wet down the treated areas for at least 48 hours.

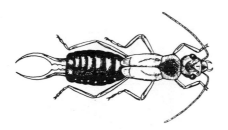

**Fig. 15. Earwig.**

## Mosquitoes

The size of mosquitoes (Fig. 16) varies from 1/4″ to 1/2″ long. This insect is slender, has long legs, and its color may vary from pale gray to black. Mosquitoes lay their eggs in water, where they hatch into larvae and live until they change into adult mosquitoes. These pests hide and breed in lawns and bushes. They attack humans and other warm-blooded animals. Some species carry diseases.

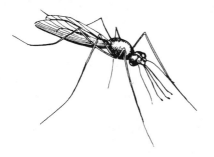

**Fig. 16. Mosquito.**

To control mosquitoes, spray all foliage, lawn surfaces, low spots that collect water, standing water, and trash areas. Do not spray any water that is to be used. Mosquitoes are easiest to control in the larvae stage. For treatment indoors, spray around doors, windows, and screens. Treat all secluded spots, such as closets, underneath furniture, behind pictures, and around screen doors.

### Scorpions

Scorpions (Fig. 17) are 2″ or 3″ long with two pincer legs in front. They resemble small lobsters and like hot, dry surroundings. The bite or sting of this insect is poisonous and causes swelling and intense pain in the joints.

*Fig. 17. Scorpion.*

To control this pest, spray or dust basements, treat all living areas, especially kitchens and bathrooms, and treat any outdoor area where scorpions are noticed.

### Silverfish

Silverfish (Fig. 18) are carrot-shaped insects about 1/3″ to 1/2″ long. They are wingless, with a silvery greenish-gray or brownish colored body, have two long antennae on their head and three similar antennae-like growths at their tail. This insect feeds on paper, paste, starched clothes, and may cause food spoilage.

269

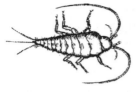

*Fig. 18. Silverfish.*

To control this pest, spray or dust all areas where they are noticed, as well as behind shelving and in storage areas.

## Spiders

The size and color of spiders (Fig. 19) varies. All spiders have four pairs of legs and two body segments. This insect spins dirt-catching webs, and some species will bite humans. The bite of the black widow can be fatal, especially to children.

To control spiders, liberally apply a wet, coarse insecticide spray to lawn surfaces, basements, trash piles, stumps, and outbuildings.

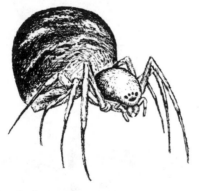

*Fig. 19. Spider.*

## Termites

Termites (Fig. 20) work under cover and feed on wood. When they cannot find sufficient quantities of wood in or on the ground, they will emerge from the soil to attack the wood in buildings, but will avoid contact with the outside air or light. Once termites reach the wood structure of a building, they work quickly and soundlessly. They will often hollow-out a board, leaving nothing but a shell of paint.

Worker termites resemble small white ants, excepting they have heavy waists and straight antennae. The presence of this destructive insect may be detected by the swarming of the reproductive members of the colony. These members are black, narrow-bodied, and resemble winged ants except that the termites have thick waists whereas ants have narrow waists.

If any wood parts of a building are in contact with the ground, termites can go directly from the soil into the wood. If masonry foundations or other "inedible" barriers are encountered, the termites may "bridge" the barrier with mud shelter tubes. These

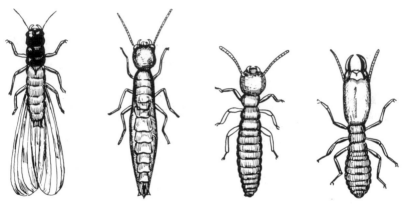

**Fig. 20. Termites.**

shelter tubes may be attached to the concrete or masonry foundation, or they may rise directly upward without support. They are a sure sign of termite infestation.

It is recommended that yards be kept clean near buildings. Remove all scrap lumber in contact with the ground and remove all tree stumps as soon as possible. Don't use waste lumber, tree trimmings, or other wood for fill, as this type of material will attract termites and sustain their colonies.

No wood portion of a building should be in contact with the ground, and all wooden posts should be set in a concrete base. If posts are equipped with metal termite shields, be sure they are not bent or damaged. Foundation cracks should be sealed as termites can pass through a 1/64" opening.

To contol termites, treat the soil around all buildings from the finished grade line to the top of the footing. Voids in masonry and beneath concrete floors may be treated by drilling. Crawl spaces beneath buildings should also be treated. It is best to have a professional pest-control operator perform these services.

## Wasps

Wasps (Fig. 21) are long, slender insects that are also known as hornets, yellow jackets, and mud daubers. They are normally about 1″ long and are either entirely black or have yellow and black markings. They will attack humans and their stings are painful and, in some instances, fatal. Their nests are usually made of

*Fig. 21. Wasp.*

clay or of a material resembling paper, depending on the variety of wasp.

To control this pest, treat the building exterior and nest by applying a spray, dust, or oil solution directly to the nest opening at night when the wasps are least active. Extreme care should be used in the procedure. Complete instructions can be found on insecticide containers.

## Powder-Post Beetles

The larvae of powder-post beetles (Fig. 22) tunnel in the woodwork of buildings and cause damage second only to that resulting from termites. The eggs of this insect are laid in or on wood, and when the larvae hatch, they tunnel into the wood, producing a fine powder or frass as they feed. After pupation, the adult beetles eat their way to the surface, making shot-holes. Many of the smaller beetles make emergency holes of about 1/8″ in diameter, and larger beetles make 1/4″ to 3/8″-diameter holes. An infestation of powder-post beetles may be detected by looking for these holes and by the small piles of ejected powdery borings.

**Fig. 22. Powder-post beetle.**

Powder-post beetles can be controlled be applying a solution of chlordane and deodorized kerosene to the surface of the infested wood. This solution may be brushed or sprayed on.

For killing deep-tunneling larvae in isolated infestations, the slow-diffusion method should be used. This procedure involves boring a downslanting hole about 1″ deep in the wood at the place where the larvae are working. A tube is inserted in the hole and a funnel fitted into the tube. The funnel is filled with chlordane solution and the equipment is left in place for 48 hours, or until the insecticide completely diffuses in the wood. If infestations are more than 18 inches apart, separate treatments must be made.

### Fleas

The flea (Fig. 23) is a small, hard-skinned, spiny insect that feeds on blood and is usually found on pet animals. Infestations have been found in basements and also in rooms inhabited by

**Fig. 23. Flea.**

pets. Outdoor locations where animals sleep often are also centers of infestations.

Spray, dust, or brush insecticide on the infested areas to control fleas. *Sevin* will rid dogs of fleas, ticks, lice, and mange mites. It may be applied once a week or every other week as the case may

273

warrant. *Do not apply sevin to young dogs or to cats.* Cats lick their fur and may ingest the powder. Insecticides malathion and baygon also help control fleas.

## House Flies

House flies (Fig. 24) are annoying pests and may also transmit disease. They are two-winged insects with four stages of development—namely the egg, larva, pupa, and adult. The adult lays its

*Fig. 24. House fly.*

eggs in animal and vegetable refuse, depositing as many as 21 egg masses, each with about 130 eggs, during its lifetime of 2 to 12 weeks. The eggs hatch in 10 to 20 hours. Because of its feeding and breeding habits, the house fly is one of the greatest insect carriers of disease. It is reported to travel 3 to 8 miles from its breeding place in search of food.

Flies can be controlled by good sanitation practices and the application of insecticides, preferably of the spray type.

## Bed Bugs

Grown bedbugs are reddish-brown, flattened, wingless insects about 3/8″ long. They lay their eggs in wall cracks, loose edges of wallpaper, furniture, bedding, and other places of protection. Adult bugs live for 5 to 8 months and may survive for as long as 7 weeks without blood food. It is estimated that a female bug may lay as many as 450 eggs during her lifetime.

Bed bugs are found in many places, such as beds, mattresses, furniture, and luggage. These pests often cling to the clothing of persons and thus can be transported great distances and brought into uninfested homes and other living quarters. They live in close association with humans.

A recommended control method for bed bugs is the use of a residual spray containing DDT. In beds, careful examination should be made of mattresses and box springs, as well as all tufting of the items. Cracks and crevices should be sprayed, as well as other locations that could possibly conceal the bugs, including baseboards and bed ledges.

## SAFETY

It is essential that every precaution be used when applying insecticides. Most are poisonous to people and to animals. Keep them out of the reach of children and pets.

Do not store insecticides with food. When applying them, be sure not to contaminate the water supply, food, dishes, or utensils.

Do not breath the dust or spray.

Follow the directions and heed all precautions on the container label.

If liquid insecticide is spilled on the skin, wash it off immediately.

When you have finished placing or applying an insecticide, wash all exposed surfaces of the body with soap and water. Wash the hands and face before eating or smoking.

Avoid spray drift into bee yards. Be careful not to get excessive insecticides into streams, lakes, or ponds.

## RODENT CONTROL

Rats and mice normally live in areas inhabited by man and often in close association with man and domestic animals. They carry diseases, contaminate foods, and may even cause fire by gnawing the insulation from electrical wires.

### Rats

The presence of rats is indicated by droppings, runways, tracks, burrows, gnawing, and nests. Fresh gnawings around doors, windows, or other building accessories are often attacked by rats

Red squill is often used as a poison for rats. It is deadly to rats but nonpoisonous to humans and other animals. Red squill is well known and readily obtainable. Several other different types of rat baits and poisons are used for control including *Warfarin* and zinc phosphide. Also available is a water-soluble anticoagulant rodenticide for poisoning water in water-scarce areas in conjunction with rat control. Information concerning rat poisons and their uses can usually be obtained from a reliable exterminator or materials dealer.

and enlarged to gain entrance to buildings. Burrows are used to gain entrance into buildings, for hiding, and for breeding and nesting. Sanitation and the proper handling of food and garbage are essential for good rat control.

Locations where rats may enter buildings should be blocked with light metal or wire mesh (1/4″ or 1/2″ hardware cloth). All openings around piping and conduits should be closed. The ordinary snap-type trap is effective against rats but requires skill and persistence. Traps with a wooden base are generally recommended and should be placed along runways or at known gnawed openings. Some baits used are bacon rinds, nuts, raw sweet potatoes, and bread dipped in bacon drippings. Peanut butter is also a possible bait.

It is essential that rat harboring materials such as lumber piles, rubbish, and other debris be cleared away.

Follow basic principles of sanitation including proper handling, storage and disposal of food supplies, refuse and waste.

Rodenticides are generally the most economical and effective means of reducing mouse and rat numbers. Take precautions in handling rodenticides avoiding food contamination. Protect humans, pets and other animals against possible poisoning.

Keep all bait materials fresh and clean. Rodents seek as much shelter as possible in their movements. Avoid placing bait in the open since those placed under and around shelters are more likely to be eaten than those baits placed in open areas.

If a trapping process is used, placement of the trap is the key to success. Traps can be placed behind obstacles, along travel routes, nailed to walls, and attached to pipes. Use good rat and mouse baits such as bacon strips, bacon scented oat meal, or a piece of fresh fish. If other animals such as squirrels, skunks, raccoon or even snakes become a nuisance contact your local conservation extension office or other Federal, state or county offices responsible for wildlife management. In many instances (barring mice and rats) local and state laws vary concerning procedures for eliminating various rodent and animal species.

## Mice

Mice contaminate foods, and shred paper and clothing for nesting materials. They also stain papers and materials with excreta, and may spread disease. Like rats, mice have runways and are noticeable outdoors during the summer time but enter buildings during the winter to escape the cold. Some house mice live indoors the entire year.

Wood-base traps work well for catching mice if used regularly. Mice are more easily trapped than rats. Mice seem to have a liking for peanut butter, and this product is often used to bait the traps. Poison baits used for rats generally work well for mice, also. Poison information can be obtained from a reliable pest exterminator or rodenticide dealer.

## FUMIGATION

Fumigation is the practice of filling the infested space with gas, dust, or vaporized liquid, to kill insects and rodents. The space can be very small, such as a trunk, or can be an entire room or building. Fumigants are generally of very small particle size. Therefore, fumigants must be used in airtight or nearly airtight spaces to prevent dissipation to other than the intended areas, and also to allow adequate exposure to the insect or rodent inhabitants.

Fumigants are extremely toxic and therefore very dangerous to humans. It is recommended that only a professional or very experienced person perform the fumigation—one who can guarantee results and safety.

# CHAPTER 11

# Heating

In those sections of the world in which the temperature falls to near freezing or below, the heating system determines the use to which the building can be put where healthful working and living conditions are concerned. When heat is provided, it is essential that every effort be made to operate the heating equipment in an economical and efficient manner.

Heating systems may generally be classified under two headings —*direct* and *indirect* types. In *indirect* heating, the heated surfaces are outside the rooms and the heated air is brought into the rooms by duct work. Air is passed over steam coils or other heat-radiating surfaces and forced into the rooms by gravity or by fans. In the *direct*-type of heating systems the heated surfaces are installed directly in the rooms or in the spaces to be heated. Steam and hot-water radiators, fireplaces, and stoves are of this type.

## HEATING TERMS

There are many common heating terms in use among those responsible for maintaining heating facilities and units. The following is a brief review of some of these terms.

*Btu (British Thermal Unit)*—A Btu is the quantity of heat required to raise the temperature of 1 pound of water 1 degree Fahrenheit.

*Specific Heat—Specific heat* is the ratio between the amount of heat required to raise a unit of weight of a substance 1 degree

F, to the amount of heat necessary to raise the same weight of water 1 degree F. One Btu is expended to raise 1 pound of water 1 degree F. However, it takes only approximately 0.115 Btu to raise 1 pound of iron 1 degree F.

*Temperature*—*Temperature* is used to indicate how hot or cold an object may be. Temperature is not heat, but the effect of heat. As an illustration, the temperature of a *gallon* of water may be 50 degrees and the temperature of a *barrel* of water also 50 degrees. Although both quantities are at the same temperature, the barrel of water has more heat than the gallon of water, and many more Btu's would be expended to raise the barrel of water from 32 degrees F. to 50 degrees F. than would be required to raise the gallon of water the same amount.

*Latent Heat of Evaporation*—In order to change water into steam after it has reached a boiling point, more heat must be applied. This is called *latent* heat (latent heat of evaporation). The number of heat units which are required to change 1 pound of water at the boiling point into 1 pound of steam is equal to 970.4 Btu at atmospheric pressure (29.921 inches of mercury at sea level or 14.7 lbs. per square inch).

*Boiler Horsepower*—The general accepted rule for calculating *boiler horsepower* is as follows: One boiler horsepower equals the evaporation of 30 pounds of water per hour from an initial temperature of 100° F. into steam at 70 pounds of gauge pressure, or its equivalent; that is, 34-1/2 pounds of water evaporated per hour from a temperature of 212° F. into steam at 212° F.

*Heat Transfer*—*Heat transfer* is accomplished by three specific methods—radiation, convection, and conduction.

*Radiation*—When heat is transferred from one body to another through a transparent medium (air, for example), it is transfer of heat by *radiation*. Heat from an open fire, heat from the sun, heat felt in a room from a radiator, etc., are all examples of radiation.

*Convection*—*Convection* is the transfer of heat caused by motion or circulation of some substance, such as liquid or gas, within a body. As an example, the hot-water boiler heating system in which the water is heated within the boiler and transferred by pipe to another section of the building.

*Conduction*—*Conduction* of heat generally occurs in a single body or a body in contact with it. A common example generally used to illustrate heat conduction is the common fireplace poker. When the end of the poker is placed in the fire, the end becomes hot and heat is transferred the entire length until the entire poker is hot.

## TYPES OF HEATING SYSTEMS

The heating systems in common use are the steam, hot-water, and hot-air types. Electric heat is being used in some locations but the costs are still above those of other heating systems.

### Steam Heating

Steam is an excellent heat distribution media as it carries a maximum amount of heat in a small volume. Steam used for heating is operated at a great range of boiler pressures. Systems operating at gauge pressures from a few psi (pounds per square inch) to 15 psi are normally termed low-pressure boilers. Those operating over 15 psi are normally said to be high-pressure boilers.

### Hot-Water Heating

Hot-water heating systems are those in which water is heated at a central source (boiler) and then circulated by pipe to convectors, unit heaters, or radiators.

In the forced-circulation system, circulation is provided by a power-driven pump. In the gravity system, water circulation depends upon the weight differential between the hot column of water going to the radiators and the cooler heavier column of water which is being returned to the radiators.

Advantages of the hot-water type of heating system are a more even heat output and, since the system is closed, the same water may be recirculated almost indefinitely when operated properly. In addition, there is no water-hammer problem as there sometimes is in steam systems.

### Hot-Air Systems

Hot-air heating includes stoves and furnaces where air is first heated by passing over heat transfer surfaces and then circulated to the location to be heated. Included in this group is the indirect warm-air system where the air is heated by a hot-water or steam coil, and then circulated by appropriate duct work to the desired space.

## COMBUSTION AND HEAT UTILIZATION

The burning of a fuel is a chemical process in which the combustible part of the fuel unites with oxygen from the air, resulting in the liberation of heat. The combustible portion of fuels is principally carbon and hydrogen, or combinations of these elements. Air is a mixture of gases, almost entirely oxygen and nitrogen. However, only the oxygen, which comprises only about one-fifth of the air by volume, takes part in the combustion. The nitrogen and other inert gases that make up the remaining four-fifths of the air are not changed in the combustion process except for an increase in temperature.

Before combustion takes place, a part of the fuel must be brought up to the ignition temperature, which is from 500° F. to 1500° F., depending on the kind of fuel being used. After combustion has started, the heat given off by the process raises the temperature of the remaining fuel to the ignition temperature, so that combustion continues.

Efficient combustion requires three conditions:

1. An adequate amount of air.
2. Proper mixing of the air with the fuel.
3. Combustion must take place in a space where the temperature is high enough to complete the burning.

Perfect combustion is that condition in which all of the combustible material is burned while supplying only the exact amount of air required to complete the reaction. In actual practice, this condition is never attained, and an excess of air beyond the theoretical requirement is supplied to insure that combustion is complete.

While excess air is necessary to insure complete combustion, the amount of excess air should be kept to a minimum, since the air not used for combustion merely dilutes the gases. This represents a direct loss, as the additional air must be heated and passed out of the vent.

## Fuel Oil

Fuels oils are commonly classified into numbered grades, each formulated for a specific requirement. These grades are described briefly as follows:

1. *No. 1*—A light fuel oil used principally in the smaller sized units for domestic heating purposes. This fuel is similar to kerosene. It is the most highly refined of the commercially available fuel oils; therefore, its price is the highest, but its heat content in Btu's per gallon is the lowest of the five grades.

2. *No. 2*—The most widely used distillate fuel oil for domestic and commercial heating use. It is particularly adaptable for dual-fuel burners where oil is a stand-by for gas.

3. *No. 3*—This grade was discontinued in 1948, but is still occasionally referred to in certain areas. The former No. 3 oil is very similar to the present No. 2 oil.

4. *No. 4*—This is the lightest of the residual fuel oils and is generally used without preheating equipment, but because of its higher viscosity, higher boiling point, and other characteristics, requires different burner equipment from the simple types of burners suitable for No. 2 oil.

5. *No. 5*—A heavy residual oil which generally requires preheating for pumping and burning. Some types of equipment will satisfactorily burn No. 5 oils of low viscosity without heating, but oils with a viscosity of 300 SSU or higher at the minimum burning temperature should have preheating equipment. This oil is frequently produced by blending heavier fuel with light oil to reduce the viscosity. Some blended oils are stable mixtures, whereas others tend to separate if stored for long periods of time. Separation of the lighter and heavier grades in storage can be troublesome, especially if inadequate heating equipment is used.

6. *No. 6*—This is the heaviest of the commercial grades of fuel oils and is generally the lowest in price per gallon

and the highest in Btu's per gallon. Preheating is essential both for pumping and burning; therefore, the original cost of the equipment and the cost of operating the burners is greater than for lighter oils. No. 6 oil is generally used in the larger installations where a skilled attendant is in charge. *Bunker C* is a term used in some cases for an oil that is equivalent to a No. 6.

Diesel-engine fuel is frequently used for heating purposes. It is available in several different grades ranging from No. 1 to No. 4 with characteristics similar to the heating oils of the same numbers. Diesel fuel can generally be burned in equipment designed for heating oils of the same number.

Fuel oils may be referred to as *distillate* or *residual*. No. 1 and No. 2 (and sometimes No. 4) are condensed from the distillation process of other petroleum products and leave practically no residue upon vaporization. No. 5 and No. 6 fuels (and sometimes No. 4) are *residual* oils and are the products that are left after the lighter fuels are distilled off. They are heavy and black in appearance and may contain residues of carbon and ash.

Fuel-oil standards, established by the American Society for Testing Materials (ASTM) and issued by the U. S. Department of Commerce, classify fuel oils into the 5 previously described numbered grades. The grade number of an oil is basically determined by the viscosity.

## LOW-PRESSURE FIREBOX BOILERS

The Kewanee Boiler factory engineers (American Standard Industrial Division) emphasize the fact that keeping a boiler clean and the firing equipment properly regulated promotes efficient operation. Here are some rules to remember:

1. Do not operate a new boiler until it has first been boiled out with a caustic solution to remove all oil and grease.
2. When starting up for the first time, fire at a low rate to dry out the refractory lining.
3. Avoid sudden heating or cooling.
4. Check water level daily.

5. Blow down the water level controls frequently.
6. Keep the boiler clean on the inside. Wash out at the end of the heating season.
7. Inspect both the fire-side and water-side surfaces for corrosion or pitting at least once each year.
8. Give the boiler immediate attention when taken out of service to prevent deterioration.
9. Never fill the boiler with water and allow it to stand without first heating the water to a steaming temperature.
10. Keep the boiler room clean, orderly, and well lighted.
11. Maintain an opening, with an area equal to that of the breeching, to the outdoors for combustion air.

Fig. 1 shows the location and identifies the tapping and trim on two typical low-pressure firebox boilers.

## PREPARING A NEW BOILER FOR SERVICE

Before operating a new boiler, it must be washed out on the inside to remove the oil and grease used in tube-rolling and threading operations during manufacture. It is essential to clean the boiler because these materials, if left in, may cause overheating and burning of the metal, foaming, and priming. The presence of even a thin film of oil on the heating surface will seriously retard heat transfer and cause blistering of the plates and loosening of the tube joints. Wash the boiler as follows:

1. Remove the safety or relief valve.
2. Close the valve in the steam or flow line.
3. Add caustic soda to the boiler water through the safety or relief valve opening at the rate of 1 lb. per thousand square feet of radiation capacity.
4. Provide a pipe connection from the safety or relief valve opening to a convenient drain to serve as a vent.
5. Fire the boiler at a low rate, but sufficient to keep the boiler water at a steaming temperature, allowing the generated steam (along with the entrained water and impurities) to discharge through the vent. With a hot-water boiler, keep the temperature at about 200 degrees.

6. Feed water to the boiler as required to maintain the water level in the glass. It may be necessary to stop the burner if the water leaves the boiler through the vent faster than it can be added through the feed line. With a hot-water boiler, keep the water trickling from the vent.

7. Continue the boiling process for 4 or 5 hours.

8. Drain the boiler while hot; then remove the manhole cover, all handhole covers, and washout plugs.

9. While the boiler is still warm, flush the interior surfaces with water from a hose nozzle under full pressure until all trace of dirt and impurities have been removed and the water coming from the boiler is clear. Start at the top of the boiler and work down.

10. Repeat the process if necessary to insure a clean boiler.

A new boiler must be fired at a low rate for at least one day to dry out the insulating and refractory linings. Adjust the burner for a low firing rate and operate intermittently, with just sufficient heat to keep the combustion chamber warm. The drying procedure may be carried out at the same time as the washing process.

## HOT-WATER BOILERS

With a hot-water heating system, the boiler and entire system must be filled with water for operation. The altitude or pressure gauge will indicate the amount of pressure required to fill the system with cold water.

## INSULATING THE BOILER

The outside surface of the boiler should be covered with insulation to prevent needless heat loss. Thus, excessive boiler room temperatures are avoided and better economy is established. The covering is usually not applied until after the boiler has been operated for a week or ten days in order to test all joints and connections for tightness. *Be careful not to cover the stampings on the boiler.* The Kewanee Boiler Company has an excellent bulletin which gives complete directions for applying insulation to a boiler. (Kewanee Bulletin No. 1014-A)

## BOILER FIRING

The boiler must be filled with water to the normal water line before operating. It is important to heat the water to steaming temperature as soon as the boiler has been filled. Do not allow the water to stand in the boiler unless it has first been steamed to

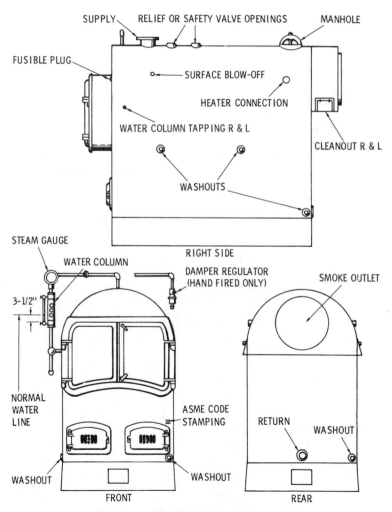

**Fig. 1. Identification and location of tapping and trim; (A) Kewanee**

drive off the air bubbles. Do not start the burner until the furnace has been purged of combustible gases from previous operation.

The firing rate should be no higher than necessary to carry the load. In general, the burner should be adjusted for a clean fire, without smoke or soot. With oil, a soft yellow fire with dark, but not smokey, flame tips is an indication of good combustion, while

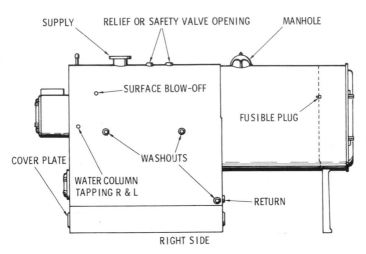

RIGHT SIDE

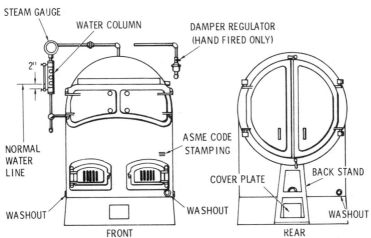

FRONT        REAR

*Courtesy American-Standard Industrial Division.*

**Type "C" boiler; (B) Kewanee 5000 Series Type boiler.**

a dazzling bright flame indicates too much air. With gas, a long purple flame indicates insufficient air. The burner service man should be consulted for any adjustment necessary on the fuel burning equipment.

### Automatic Firing

If a boiler is fired with either an oil, gas, or combination oil and gas burner or stoker, and is equipped with all controls and safety devices, operate the burner in accordance with the burner manufacturer's directions. These are obtainable either from the heating contractor who made the installation or directly from the burner manufacturer.

### Hand Firing

**Starting Fire**—Cover the grates with coal to a depth of about 6 inches, using small sizes of coal. Place wood kindling and paper at the rear of the grates and ignite (Fig. 2). Fig. 3 shows the condition of the firebed when refueling is needed.

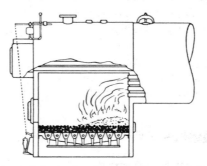

*Fig. 2. View of the boiler firebox when starting a fire.*

*Courtesy American-Standard Industrial Division.*

**Coking Method**—Use the coking method of firing for high efficiency and minimum smoke. Push the live coals to the rear of the grates with a hoe, leaving the front of the grates nearly bare, as in Fig. 4. Pile the live coals high enough so they will not be covered over by the fresh coal. Then place the fresh fuel charge over the front portion of the grates and against the live coals, as in Fig. 5, but do not cover the live coals. The combustible gases mixed with the over-fire air will burn when ignited by the live

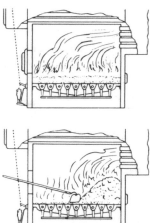

Fig. 3. Condition of the firebox when refueling is needed.

Fig. 4. Pushing live coals to the back of the firebox prior to refueling.

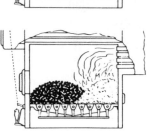

Fig. 5. Fresh fuel should be added to the front of the firebox when refueling.

*Courtesy American-Standard Industrial Division.*

coals, thereby producing a hotter fire with little smoke and soot. Provide over-fire air through openings in the firedoor.

Carry as deep a fuel bed as the draft will permit. Do not allow the fire to burn too low before refueling. The amount of live coals to be pushed back should about equal the amount of fresh coal to be added.

**Alternate Method**—This method may be used instead of the coking method. First, push all live coals to one side of the grates, then place fresh coal on the side opposite the live coals. Continue to place fresh coal on alternate sides each time coal is added.

**Spread Method**—For high steaming rates, this method may be desirable. Fire with small quantities of coal and on alternate sides each time coal is added.

**Banking Fire**—When fires are banked and the furnace temperature reduced, make sure to have enough draft through the boiler to prevent the accumulation of combustible gases within the setting.

**Draft Control**—The best results are obtained by maintaining a low, steady draft. The draft adjuster should be set to keep the draft as low as possible without having combustion gases leak out into the boiler room.

**Ashes**—Remove ashes often. Do not allow the accumulation of any large amount of ashes in the ashpit, boiler flues, chimney connections, or chimney base.

## Water Controls

Always note the water level in the glass and the steam pressure on the gauge when entering the boiler room. Observe the reading of the steam gauge in relation with the setting of the pressure control.

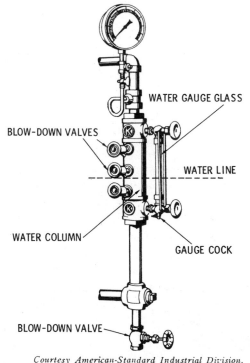

*Courtesy American-Standard Industrial Division.*

**Fig. 6. A typical water control installation.**

The gauge cock and the drain valves on the water column and water glass should be operated daily to make sure these connections are clear. Neglect of this duty could allow the connections to become choked with sediment so that the true water level would not appear in the glass. Fig. 6 shows a typical water control installation.

Blow out the low-water cutoff and water feeder as recommended by the manufacturer. Test the control by turning off the water supply and then observing whether or not the burner cuts out with low water in the glass. With a combined boiler feeder and a low-water cutoff, the feeder valve closes with 1 inch of water in the glass. The cutoff switch stops the burner when the water level drops to 1/4 inch in the glass, and when the water level is restored to 3/4 inch in the glass, the switch starts the burner.

## BLOWOFF VALVE

The use of the blowoff or drain valve in the low-pressure heating boiler is for the purpose of discharging rusty water and sediment which may settle to the bottom of the boiler. The withdrawal of a pail of water about once each month is generally satisfactory.

## WATER TREATMENT

Water treatment in a low-pressure heating boiler is mainly for protection from corrosion or pitting. There is usually no lime or scale problem, as the same water may be used over and over again. It is recommended that the boiler water be treated for corrosion. Initial treatment at the beginning of the heating season is generally satisfactory for the entire year, as long as the same water is used.

## FOAMING AND PRIMING

Foaming or priming in the steam boiler will result in large quantities of water being carried over into the steam main. This condition will cause violent fluctuation or a sudden disappearance of the water in the glass. Impurities will show in the boiler water. The following reasons may be the cause of trouble:

1. Dirt or oil in the boiler water.
2. Overdose of boiler compounds.
3. Carrying too high a water level in the boiler.
4. High overload on the boiler.

In case of serious trouble, stop the burner and decrease the load on the boiler until the true water level can be determined. Then alternately blow down and feed fresh water several times. If the trouble continues, it will be necessary to wash out the boiler and refill with fresh water. Blow out the safety valve and the connections to the pressure gauge, water column, and water glass to make sure that they are clear and unobstructed by the impurities which occur with priming.

## CHIMNEY MAINTENANCE

### Inspection

Inspect chimneys at least once a year, preferably in the fall. Inspect for loose or fallen bricks, damaged flue lining, and excessive soot accumulation. Inspect interior of chimneys by lowering an electric light into the chimney. Outside chimney mortar joints can be checked by probing with a knife or screwdriver.

If defects are found make necessary repairs. It is preferable not to use old soot or creosote impregnated brick in repair work because these may leave a stain on interior plaster or wallboard if excessive dampness occurs.

### Cleaning Chimneys

Chimney cleaning is not often required but if excessive soot accumulation is noticed cleaning is required. Hiring a commercial cleaning firm is the best procedure. However, if work is to be done personally, soot and loose material can be removed by pulling a weighted sack of straw on a rope up and down the chimney flue. If a fireplace exists be sure to seal off the fireplace opening to prevent soot and fly ash from getting into the building interior. A fireplace opening can be sealed using a plastic drop cloth and masking tape.

Common rock salt can be used as a chimney soot remover although it is not always effective. If salt is used place about two

(2) teacupfuls in the fireplace or flame location when a good fire is burning. Note! Chemical commercial soot removers are not very effective in removing soot and creosote and may cause the soot to burn creating a fire hazard.

## Chimney Spark Arrestors

Chimney spark arrestors may be required when chimneys are near combustible roofs, woodlands, or other combustible material particularly if sawdust or paper trash is burned. When building an arrestor a rust-resistant screen should be used with screen openings not larger than ⅝″ (five-eighths inch) or smaller than ¼″ (one-quarter inch). Fasten screens securely to the top of the chimney. Spark arrestors are not recommended if soft coal is burned since the openings may clog with soot. Inspect fire arrestors for deterioration and clogged openings at least once a year.

## FLUE CLEANING

The boiler flues should be examined regularly and scraped clean if they are coated with carbon, soot, or fly ash. More frequent cleaning will be required with a coal-burning unit than with one burning oil or gas.

## MINOR REPAIRS

### Piping Leaks

All piping connections to the boiler and all accessories should be maintained in a leakproof manner, as even a minor leak will soon become serious if neglected. This applies particularly to the water-column, water-glass, water-level control piping, and manhole and handhole gaskets. Pipe nipples should be tightened or replaced if there is evidence of leakage. Leaky plugs in the boiler shell should be replaced with nipples and caps.

### Replacing Water Glass

The water glass should be removed and cleaned as often as necessary. When a water glass has been broken, remove the broken pieces and slowly open the valves to blow out any remaining pieces. Before inserting the new glass, see that the drain is open and that the replacement glass is of the exact length required and that all connections are in line. Insert the glass, rubber gaskets, and brass rings, and tighten the stuffing-box nuts, taking care not to set the nuts too tight. Warm the glass by opening the top valve slightly and letting a small current of steam flow through the glass. Close the drain cock after the glass is sufficiently warmed, open the bottom valve slightly, and when the water level in the glass has become stable, open the bottom valve wide and then open the top valve wide.

## Applying Manhole and Handhole Covers and Plugs

Use new gaskets when replacing manhole and handhole covers and plugs. Clean the metal surfaces where the cover plate bears against the shell plate. Apply graphite paste or grease to the gasket to prevent sticking and to secure tightness. Use care in centering the cover plate and gasket in the shell opening before drawing up the bolts. Draw the bolts up firmly and, if necessary, tighten them repeatedly when raising steam pressure to make them leakproof. Grease the washout plug threads and use new gaskets.

## LOW-PRESSURE BOILER MAINTENANCE

In placing a boiler out of service, it is best that it be kept clean and dry, both internally and externally. Hot-water boilers and systems may be left filled with water during shut downs, providing the boiler room is dry. However, condensation and rust is less likely to occur if the boiler is drained and kept dry.

### Washing Boiler

To wash a boiler, remove the handhole and manhole covers. Hose the inside of the boiler with water under high pressure, and use hand scrapers to remove any mud, scale, etc. Always start at the top and work down.

Dry the boiler by using hot-air mechanisms, stoves, etc., or by building a small fire of paper in the firebox. Do not let the boiler get too warm to be uncomfortable to hand touch. Leave all openings uncovered during the drying process. If the boiler is gas fired, leaving the pilot burning will generally keep the boiler dry.

### Cleaning the Outside of the Boiler

Completely scrape all carbon and soot from the fire surfaces. After the fire tubes have been cleaned, swab them out with a mineral oil, such as crankcase drainings. Remove any rust or deposits from the shell, and after completely cleaning, paint with a coating of rust-inhibiting paint, such as red lead. Make sure no water or steam can enter the boiler. When starting up the boiler after a long shut down, fill the boiler with water, wash out, and refill.

## BOILER OPERATING LAWS AND ORDINANCES

Boiler operation is strictly regulated in some localities and cities. It is therefore recommended that the boiler operator be familiar with all local laws relating to the duties of the engineer and fireman, or to the safety requirements of his work.

## INSPECTIONS

All boilers installed in those states where periodical inspection is the law will be inspected by the State or Municipal Inspector having jurisdiction and authority. It is recommended that these inspections be supplemented with inspections by the person responsible for the particular heating facility or plant.

## GENERAL BOILER MAINTENANCE
### (High- and Low-Pressure)

1. If oil-fired, keep a continuous check on the fuel supply. Do not depend on "automatic delivery"—watch the supply yourself. Check the fuel gauges on the oil tank. Keep track of the supply and re-order before the tank is empty.
2. Keep the boiler, burner, and the entire boiler room clean. Do not allow fuel to leak anywhere—*it is dangerous*. A clean boiler room is essential to first-class boiler operation.
3. Keep the burner control cabinet door closed. The electrical contacts in the cabinet are very sensitive to dust and dirt.

4. Never close the vents supplying air to the boiler room. If cold air currents cause difficulty with other boiler-room equipment, the air ducts should be installed to direct the flow of fresh air.

5. Repair all leaks promptly. All piping connections to the boiler and accessories should be maintained leak-proof because even a minor leak, if neglected, may soon become serious. This applies especially to the water column, water glass, water-level control piping, and manhole and hand-hole gaskets. If serious leaks do occur, shut down the boiler immediately and gradually reduce the steam pressure. *Do not attempt to make repairs while the boiler is under pressure.*

6. Any problem in regard to extreme foaming or priming, scale in the boiler, corrosion, or pitting should be referred to a concern specializing in boiler-water chemistry. *Do not experiment with "homemade" treatments or compounds.*

7. *A boiler filled with raw water should not be left idle.* The water should first be boiled with the steam chamber vented to drive off any dissolved gases. In the case of hot-water boilers, heat the water in the boiler to 200° for the same reason. The boiler water should then be made alkaline with caustic soda. Use 1.5 ounces of caustic soda per 100 gallons of water contained within the boiler.

8. *Be careful*—Do not add large quantities of cold feed water to a hot boiler.

## Daily Maintenance

1. Check the boiler water level in the glass and the steam pressure on steam boilers. Check the temperature reading and water pressure on hot-water boilers.

2. On high-pressure steam boilers, blow down with at least one full opening and closing of the blow-down valve, except where the amount and frequency of blowing down are determined by chemical analysis.

3. On high-pressure steam boilers, the gauge cocks and the blow-down valves on the low-water cutoff, water column, and water glass should be operated to make sure these connections are clear.

## Weekly Maintenance

1. Check the boiler flues and clean if required. The frequency of flue cleaning will depend on the fuel used, the burner setting, and other factors of the job performance. Protect the burner and controls from dust and dirt during the tube cleaning operation. Make sure that the linkage and other light parts are not damaged.
2. Inspect all burner linkages on oil-burner units and tighten as required.
3. The low-water cutoff should be tested by turning off the water supply, opening the blow-down valve to remove rust and dirt, and then noting whether or not the burner cuts out with low water in the glass.
4. On low-pressure steam boilers, the gauge cocks and blow-down valves on the water column and water glass should be operated to make sure these connections are clear.

## Monthly Maintenance

1. Check the ignition assembly and electrodes on oil-burning units. Clean if necessary.
2. Clean the oil-atomizing nozzle if necessary. *Never use a sharp instrument on the nozzle.* If the nozzle becomes damaged, be sure to replace it.
3. Clean the scanner.
4. Check the primary and secondary air dampers. Remove any accumulation of lint or dirt.
5. Inspect the condition of the refractory material.
6. Inspect the oil filter and replace the cartridge if necessary. The frequency of cleaning the oil filters will depend upon the quality of the oil used.
7. Lubricate all motors as recommended by the motor manufacturer's instructions.
8. On low-pressure boilers, drain off a small amount of water from the boiler drain valve to remove the sediment and rust accumulation. *Don't overdo and waste clean water.*
9. Test the boiler safety valves.

## Annual Maintenance

1. The burner should be inspected and checked by a competent service man. If the unit is fired by oil, field replace-

ment of the internal parts of any of the pumps, such as the seals, gears, or shaft, is discouraged. A factory overhaul or a replacement of the complete pump assembly is suggested since failure of one part usually indicates failure in other parts.

2. Replace the vacuum tubes and the scanner cell in electronic controls. This replacement is inexpensive insurance against future service calls.

3. Check the condition of the oil tank. Clean and remove sludge if necessary.

4. If the boiler is to be out of service for the summer, be sure to break all power connections to the boiler and auxiliaries.

5. To protect the fire side of the boiler from corrosive action of combustion deposits, scrape all carbon, rust, soot, or other deposits from the fire surfaces. Apply a thin coating of oil or grease if the boiler is to remain out of service for the summer.

6. To clean the water side of the boiler, remove all manhole and handhole covers. Hose the inside of the boiler with water under high pressure, and use hand scrapers to remove any mud, scale, etc. Start at the top of the boiler and work down.

7. Call for a boiler inspection, if required by your state, municipal, or insurance agencies. The boiler must be fully accessible both inside and out for the inspector.

8. Since local conditions determine the use of wet or dry storage during summer shut-down periods, information as to the type of boiler protection required should be obtained from your boiler-water consultant or insurance company.

9. *Wet Storage.* Be sure that the feed-water and steam connections are tight. After a light steaming period vented to the atmosphere to expel dissolved gases, completely fill the boiler, and add sodium sulfite (approximately 1.5 lbs. per 100 gallons of water within the boiler) to provide a protective concentration of chemicals. Tests should be conducted occasionally to make sure that the boiler is properly protected.

10. *Dry Storage.* Drain the boiler, clean it internally, and then dry it thoroughly. In damp boiler rooms, place trays of moisture absorber, such as quicklime or silica gel, inside the boiler shell. Close all openings to exclude moisture and air.

## BOILER AND FURNACE BRICKWORK

Some large industrial furnaces and boilers require a refractory lining. These linings are designed and constructed for the most efficient service. To utilize good design, careful attention to refractory construction is required. Such construction is a highly specialized and skilled trade and should be entrusted only to experienced tradesmen.

To operate efficiently, this type of furnace wall should be constructed to meet the following requirements, as far as is practical:

1. The strongest possible bond or tie should be provided across joints and through the wall.
2. The construction should be as air tight as possible.
3. Provision should be made for removing and rebuilding the inside brick courses on the hot face as easily as possible and with the least disturbance to the rest of the wall.
4. Provision should be made for relieving the fluctuating expansion strains by allowing proper expansion joints.

The determining factor in meeting the first three requirements is the type of bond used and the soundness of the brick laying. The type of expansion joint is not dependent on the kind of bond used.

Fig. 7 illustrates some of the standard masonry bonds found in boiler and furnace construction. The wall thickness and type of bond is determined by the service conditions present and by the load due to the height of the wall. To achieve maximum strength plus an airtight or gastight wall, it is important to stagger the vertical joints. The alternate header and stretcher bond is good construction practice and is probably the most common.

### Expansion Joints

One of the important factors to consider in designing and constructing furnaces is that the brickwork expands and contracts

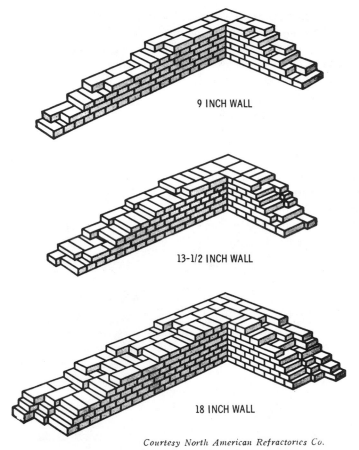

9 INCH WALL

13-1/2 INCH WALL

18 INCH WALL

*Courtesy North American Refractories Co.*

**Fig. 7. Typical masonry bonds used in boiler and furnace construction.**

due to heating and cooling. Severe fluctuating stresses are often developed as the temperature inside the furnace rises and falls. This is supplemented by the steep temperature gradient usually present in furnaces whereby the hot face expands more than the cooler parts of the brickwork. It should be noted that this expansion takes place both horizontally and vertically.

Provision for vertical expansion, except in high walls, is easily made by leaving the wall free to expand upward without encountering boiler drums, piping, or other structural members. Vertical contraction during subsequent cooling takes care of itself when the wall settles.

There are two common types of expansion joints for horizontal movement, as shown in Fig. 8. The *staggered* or *broken* type has the open joints staggered at adjoining courses. This type is perhaps the simplest and easiest to construct as it requires no large 9-inch straight brick, or any cutting and fitting. This type also results in a minimum weakening of the wall at the expansion joints. The *straight* or *vertical* type with no overlapping headers allows the wall sections to expand and contract with a greater freedom which results in a minimum of buckling on the inside brick courses. As the ability to open and close easily is the expansion joint's main purpose, the free movement of the *straight* type probably outweighs its disadvantages.

STRAIGHT OR VERTICAL TYPE

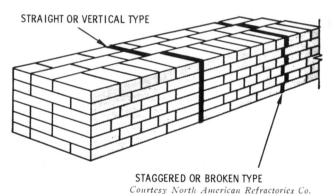

STAGGERED OR BROKEN TYPE
*Courtesy North American Refractories Co.*

**Fig. 8. Two types of expansion joints to provide for horizontal movement of the brickwork as it is heated.**

In practice, expansion joints are usually constructed to provide for twice the expansion that will occur on the first heating. To keep joints clean, they are usually packed with a combustible or compressible filler during construction.

## STEAM TRAPS

All steam traps should be disassembled at least once a year to check the condition of the operating mechanism. The following information is listed to aid in the maintenance and repair of these units.

### Valves and Seats

If the valve seat has a sharp smooth edge, and if there is a narrow bright ring all the way around the ball valve, chances are that the valve is tight. Valves and seats that have become wire drawn or badly grooved from wear should be replaced. *Do not use a new seat with an old valve or vice-versa.* Valves and seats are factory-lapped together in matched sets for a perfect fit.

*Valve-Seat Installation*—When installing valve seats in which the joint is made, not by the threads, but rather by the contact between the ground end of the valve seat and the beveled seating area at the bottom of the tapped hole, *do not use any pipe dope or lubricant on the threads.* As shown in Fig. 9, make sure that the seating area is perfectly clean before screwing the seat into position.

METAL TO METAL JOINT MADE HERE

**Fig. 9. With certain types of valve seats, it is important that the seal is made at the point of contact shown, and not by the threads.**

*Courtesy Armstrong Machine Works.*

*Valve Lever*—When a valve and seat need replacement, the fulcrum points on the valve lever usually will be worn enough to reduce the bucket travel and the trap capacity. Replace the valve lever and valve retainer at the same time the valve and seat are replaced and the trap will be like new. Be sure the new valve parts are the right size for the trap and the operating pressure.

The fulcrum points on the valve lever are carefully adjusted for height before the lever leaves the factory. If the fulcrum points, guide pin holes, or bucket hook slot are excessively worn, replace with a new lever having same size stamping as the valve seat.

*Alignment of Guide Pins*—To check the alignment of the guide pins, hold the lever assembly against the valve seat with the valve contacting its seat, and the two fulcrum points resting on the face of the seat. When the lever is held in this position, the guide pins should be centered in the guide pin holes, as shown in Fig. 10A.

There should be equal side-to-side movement of the lever, as shown in Figs. 10B and 10C. It is easy to bend the pins until they are

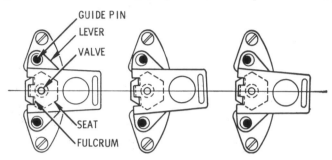

GUIDE PIN
LEVER
VALVE
SEAT
FULCRUM

*Courtesy Armstrong Machine Works.*

**Fig. 10. Guide-pin alignment. When the guide pins are positioned as in A, the lever can be moved sideways the same distance to the left (B) as to the right (C).**

centrally located. Care should be taken to keep the pins perpendicular to the guide-pin plate so that the lever can drop until it rests on the guide-pin hooks. Two examples of *incorrect* alignment are shown in Figs. 11A and 11B.

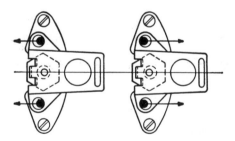

*Courtesy Armstrong Machine Works.*

**Fig. 11. Two examples of incorrect guide-pin alignment. The pins should be bent in the direction of the arrows until they center in the holes.**

### The Lever Assembly

The lever assembly is hooked over the guide pins. In a few sizes of traps, particularly at low pressures, the valve-lever assembly must be slipped on the guide pins, as shown in Fig. 12, before the guide-pin assembly is fastened into position.

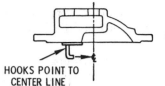

HOOKS POINT TO
CENTER LINE

*Fig. 12. The guide-pin hooks are installed pointing toward the centerline of the trap.*

### Guide-Pin Assembly

Replace this assembly if the pins are badly worn or grooved. Be sure the size stamping on the new assembly is the same as on the lever and orifice. This part should be installed with the guide-pin hooks facing the center of the cap.

### Buckets

Cracked or corroded buckets should be replaced. Hold thermic-type buckets over a steam jet or a lighted match to see that the disc seats properly when the bimetal strip is heated. These thermic strips are replaceable.

### Dirt In Trap

Remove all sediment and other dirt from the trap body. The mechanism may require cleaning by immersing in a solvent, gasoline, or kerosene. *Use extreme care as these liquids are flammable.* If there is an exceptional amount of dirt, install a strainer ahead of the trap. The strainer will have to be blown down or cleaned at periodic intervals.

### Bypass Valve Inspection

If traps are installed with a bypass, it is highly important that the bypass valve be checked to make sure it is perfectly steam tight. If the trap can be operated without the bypass, it should be removed. Avoid the practice of opening bypass valves and leaving them open.

### Check-Valve Inspection

Make sure that check valves ahead of the trap, in the trap, or in the return line, are tight and in good condition.

### Pressure Changes

Having had experience with Armstrong traps, it has been found that these will operate at any pressure lower than the maximum for which they are furnished. The maximum pressure depends upon the diameter of the discharge orifice used in each size of trap.

If it is necessary to change the working pressure of the trap to obtain greater capacity at lower pressures, or to enable the trap to work at higher pressures, a complete pressure-change assembly is required. This comprises a valve seat, valve, valve retainer, valve lever, and a guide-pin assembly. The diameter of the valve seat is stamped on the face of the seat itself, on the valve lever, and on the guide pin-assembly. Parts having different stampings should never be used together.

## Unit Heaters

Unit heating systems are, on the average, economical and perform well. However, even as with other heating systems, job analysis and unit selections must be made carefully. Unit heaters offer the advantage of less temperature differentials per foot of elevation on the interior of the building than direct radiation of the wall-hung and free-standing type.

Unit heaters are designed to maintain predetermined temperatures in a given area. In order to accomplish this, their capacity must be sufficient to offset the losses due to cold-air infiltration through cracks around windows and doors, and to offset the temperature differences between the inside and outside of the facility. The building size, type of construction, temperature differentials, and infiltration of air affect the number, type, and capacity of the heaters selected.

A unit heater can deliver its manufactured rated capacity only if it is properly installed with well-designed piping systems, and if it is properly maintained. Based on engineering experience, the following piping suggestions are recommended:

1. Install the units level and plumb.
2. Suspend the heaters securely, with provisions for simple removal.
3. Provide enough work space around the units for convenient cleaning and inspection.
4. Provide for proper drainage of all steam and return lines, and for proper drainage of the condensate at the end of the steam main or line.

5. Do not drain the condensate from the steam main through a unit heater.

6. Provide for expansion and contraction of the piping that connects each unit.

Typical unit-heater piping diagrams are shown in Figs. 13 and 14.

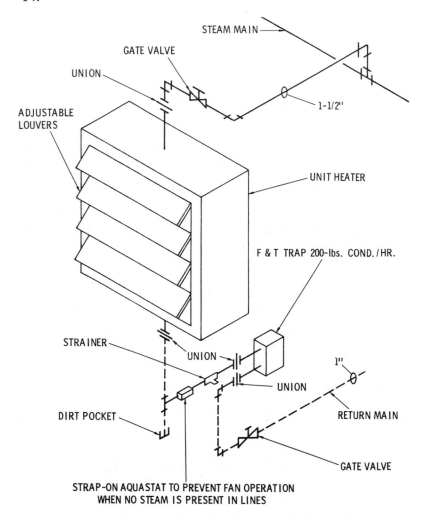

STEAM MAIN

GATE VALVE

UNION

1-1/2"

ADJUSTABLE LOUVERS

UNIT HEATER

F & T TRAP 200-lbs. COND./HR.

STRAINER

UNION

UNION

1"

DIRT POCKET

RETURN MAIN

GATE VALVE

STRAP-ON AQUASTAT TO PREVENT FAN OPERATION
WHEN NO STEAM IS PRESENT IN LINES

**Fig. 13. A typical horizontal-discharge piping diagram for a unit heater.**

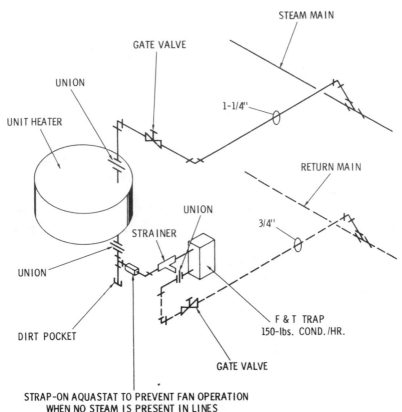

STEAM MAIN

GATE VALVE

UNION

1-1/4"

UNIT HEATER

RETURN MAIN

UNION

STRAINER

3/4"

UNION

DIRT POCKET

F & T TRAP
150-lbs. COND./HR.

GATE VALVE

STRAP-ON AQUASTAT TO PREVENT FAN OPERATION
WHEN NO STEAM IS PRESENT IN LINES

*Fig. 14. A typical vertical-discharge piping diagram for a unit heater.*

## Maintenance (Steam and Hot-Water Units)

In order to provide longer life and overall economical operation and efficiency, regular maintenance inspections of unit heaters are required. A regular card index file should be made on each unit heater, or a card on every building and/or every floor, as conditions may dictate.

Experience has indicated that the cycle of inspection is determined by the location of the heater. In locations where soot, dust, or corrosive fumes are encountered, inspections should be made every 2 months; in average locations where excessive dust or other foreign materials are not encountered, every 4 to 5 months.

Specific inspections points should include pipe-line connections, motors, fans, casings, vents, and heating elements.

*Internal Corrosion*—In order to protect the entire heating system, including unit heaters, every effort must be made to prevent internal corrosion. The boiler feed water should be treated with the proper boiler compound as suggested by the boiler manufacturer or a reputable compound manufacturer. Care must be taken in selecting the compound so as not to use an unsuitable material or an excessive amount. When large amounts of make-up water are used, it has been found that de-aerating the boiler feed water is a good practice. Check all traps for leakage or sticking, clean all dirt trap connections, and clean the strainers ahead of the traps.

*Motors*—Motors should be lubricated in accordance with the manufacturer's instructions. Many of the smaller motors (up to 1/6 hp) are equipped with sealed bearings which eliminate the need for periodic lubrication for the life of the motor. Many larger motors, particularly on horizontal units, are of the sleeve-bearing type with oil cups. These require lubrication at about 4-month intervals. A standard #10 SAE oil may be used to lubricate these motors. Some larger motors, particularly on down-blast units, have ball bearings of the grease-lubricated type designed for long periods of operation (at least 1-year) before requiring relubricating. Some motors are fitted for a pressure gun. For these, use a high-temperature grease or one recommended by the manufacturer. In all instances, be careful not to over-lubricate.

*Casings*—All dirt, both on the inside and outside, should be wiped off the casing of the unit heaters. Check all fastenings for looseness or wear—looseness may result in vibration noise. If excessive rust spots are present, sand and spot paint them with a matching color metal paint or enamel, as may be required.

*Fans*—Fans should be wiped clean, set screws inspected for tightness, and blades for possible cracks, looseness, and vibration. The fan assembly must be properly positioned on the fan shaft. The set screws and/or keys must be tight.

*Hot-Water Units*—If possible, it is a good policy to zone or section hot-water systems so that each unit can be drained individually. This simplifies the draining problems. If the unit is connected to an outside air supply and is shut down in cold weather, the system must be drained to prevent freezing and damage. Each unit should be provided with air valves of a reputable manufacturer for the purpose of complete and rapid air removal.

*Heating Coils*—Foreign matter, such as dirt, grease, lint, etc., should be cleaned from the heating elements, as such material can seriously reduce the heating capacity and efficiency of the unit.

These elements should receive close attention during the standard inspection cycle and be cleaned as may be necessary. The maximum allowable cleaning cycle should not exceed 10 months. Cleaning can be accomplished with a brush or air hose. If an air hose is used, care should be taken not to blow the accumulated dust and dirt into the motor housing.

## AIR FILTERS

The most common air filters in use are the viscous type which may be either washable, automatic, or throwaway. This type of filter cleans the air by trapping the impurities (mostly dust and lint) against the surfaces which are coated with viscous material. Removable units have metal screening which can be cleaned by recoating when required. Exhaust hoods in food preparation locations and other similar facilities normally are equipped with this type.

Renewable or throwaway filter units are made of a less expensive material, such as glass wool, and these work well. The efficiency of this type is adequate for general air filtration, and the operating cost is less than with other types.

Dry filters use a collection surface of special paper, cloth, or some similar material. The surfaces may be cleaned several times, but eventually these also must be replaced. This type is used where exceptional high efficiency is required.

Some air-conditioning systems use a water spray for air filtering. The air washer is located in a special chamber into which the air is diverted. Here, the air is brought into contact with water sprays and is washed before recirculation. The water spray drops into a pan at the base of the chamber and is recirculated by a pump for reuse. Eliminator plates are located at the air-discharge connection so that the water spray is not projected into the outgoing air stream. Experience has indicated that this type of system does not work as well as the viscous type, particularly where greasy dust particles exist.

Electrostatic precipitators are sometimes used to clean the air. These precipitators are cleaned by washing to remove the collected dust. Some maintenance men have said they have had good results in cleaning the dust away by lightly rapping the accumulator plates.

In the electrostatic precipitation method of cleaning, dust and other foreign particles receive an electric charge while passing through an electrostatic field, and then are collected on metal plates having an opposite polarity. Although this type of air cleaning is more expensive than other methods, its cleaning efficiency is very high. The method is used with effectiveness where it is necessary to remove smoke or very fine particles of dust.

## Maintenance

It is essential that air filters be serviced at regular intervals. Dirty filters result in high operating costs, poor ventilation and air distribution, and have resulted in equipment failure.

*Cleanable Filters (Steel or Aluminum Mesh)*—This type filter can be renewed by washing in a strong solvent, such as trisodium phosphate. (Rubber gloves should be worn by servicing personnel.) After washing the filters, allow them to dry completely, and then recoat by dipping or spraying with an adhesive. Filter manufacturers generally have available a spray adhesive which they recommend for their particular product. It should be odorless, fire resistant, evaporate slowly, and prevent mold formation.

*Throwaway Filters*—Washing and oiling this type of filter is not recommended. They can sometimes be cleaned by tapping an edge to the floor. However, it is best to establish a replacement schedule which can be determined by experience. They can be roughly checked by holding up to a light—if little or no light shines through, they must be replaced. The general practice is to replace the filters at least three times per year in accordance with an experience schedule. However, conditions may warrant replacement more or less frequently. It is essential that the filters be installed in accordance with the manufacturer's recommended direction of air flow. Arrows on the filter generally indicate which direction the air should flow through the unit.

## GAS-FIRED HOT-WATER HEATERS

The average hot-water heater is a quality product designed for a long and trouble-free life. The lighting instructions for this type of heater are usually found on a decal attached to the front of the unit. *Follow them.*

The thermostat on hot-water heaters is usually set at 140° F. for normal use. If hotter water is desired, such as for automatic dishwashers, washing machines, etc., the thermostat can be set higher, but should not exceed 160° F. All hot-water heaters should be equipped with a safety valve.

A hot-water heater requires little maintenance, but will last longer with less interior lime deposits if some of the water is drained off at least once a month. Draw out several pailfuls of water from the drain valve at the bottom of the heater. Dust and lint around the pilot may result in some trouble. Keep this area clean. Do not store brooms, mops, and similar items near the hot-water heater.

## INDUSTRIAL OIL-BURNER TROUBLE CHART

It is suggested that all mechanical equipment be serviced and maintained in accordance with the manufacturer's recommendation, as listed in the service manual. The following symptoms, their possible cause, and possible remedies are included as a handy reference for the most common troubles.

| Symptom and Possible Cause | Possible Remedy |
|---|---|
| **No Oil Delivery** | |
| (a) Pump not primed. | (a) Bleed any air from pump and reprime. |
| (b) Suction lift too high. | (b) Fill tank or raise to reduce lift. |
| (c) Air leak in suction line. | (c) Tighten fittings or replace defective line. |
| (d) Wrong direction of pump rotation. | (d) Reverse pump motor rotation or replace with one having the correct rotation. |

| *Symptom and Possible Cause* | *Possible Remedy* |
|---|---|
| (e) Pump coupling improperly installed. | (e) Install properly. |
| (f) Pump gears worn. | (f) Replace gears or install new pump unit. |
| (g) Pump seal leak. | (g) Tighten or replace seal. |
| (h) Pump motor not operating. | (h) Replace blown fuses, repair electrical defect, or replace motor. |

## Capacity Too Low

| | |
|---|---|
| (a) Suction lift too high. | (a) Fill tank or raise to reduce lift. |
| (b) Air leak in suction line. | (b) Tighten fitting or replace defective line. |
| (c) Suction line too small. | (c) Replace with larger line. |
| (d) Check valve or strainer too small or obstructed. | (d) Clear obstruction or replace unit with one that is larger. |
| (e) Worn pump. | (e) Repair or replace pump as required. |
| (f) Defective seal. | (f) Replace seal. |
| (g) Pump coupling slipping on shaft. | (g) Tighten coupling or replace. |

## Noisy Pump

| | |
|---|---|
| (a) Pump not securely mounted. | (a) Secure pump properly. |

| *Symptom and Possible Cause* | *Possible Remedy* |
|---|---|
| (b) Vibration due to misaligned or bent shaft. | (b) Align or replace shaft as required. |
| (c) Pump overloaded. | (c) Install larger pump or correct cause of overload. |
| (d) Air leak in suction line. | (d) Tighten fittings or replace defective line. |
| (e) Suction lift too high. | (e) Fill tank or raise to reduce lift. |

**Leaky Pump**

| | |
|---|---|
| (a) Cover bolts loose. | (a) Tighten bolts. |
| (b) Cover gasket broken or defective. | (b) Replace gasket. |
| (c) Pump seal scratched due to dirt. | (c) Replace seal. Check and replace oil filter if necessary. |
| (d) Pump bushings and other parts badly worn due to abrasives in fuel oil. | (d) Repair or replace pump. Check and replace oil filter if necessary. Check fuel oil for abrasive content. Change brands if necessary. |
| (e) Oil-line fittings loose. | (e) Tighten or replace fittings. |

## DOMESTIC OIL-BURNER TROUBLE CHART

**Burner Fails to Light**

| | |
|---|---|
| (a) Thermostat setting is lower than temperature of room. | (a) Set thermostat higher to check burner operation. |

| Symptom and Possible Cause | Possible Remedy |
|---|---|
| (b) Furnace switch turned off. | (b) Turn switch on. |
| (c) Fuses blown. | (c) Replace fuses. |
| (d) Out of oil. | (d) Fill tank. |
| (e) Fuel strainers clogged. | (e) Clean or replace strainers. |

### Raw and Stringy Flame

| | |
|---|---|
| (a) Partially plugged nozzle. | (a) Install new nozzle. |
| (b) Air in pump. | (b) Bleed air from pump. |
| (c) Incorrect air adjustment. | (c) Adjust volume of air for proper flame. |

### Ignition Points Collect Carbon

| | |
|---|---|
| (a) Nozzle loose in gun. | (a) Tighten nozzle. |
| (b) Improper oil cutoff when burner shuts down. | (b) Check pressure regulator valve and clean if necessary. |
| (c) Ignition points too close to nozzle. | (c) Adjust according to service manual. |

### Noisy Pump

| | |
|---|---|
| (a) Excessive suction on oil line due to a possible plugged strainer. | (a) Clean or change filter. |
| (b) Water condensed in oil storage tank. | (b) Drain or siphon water from tank. |
| (c) Leaks in suction line. | (c) Correct leakage. |
| (d) Air in oil line or pump. | (d) Bleed air from pump. |

*Symptom and Possible Cause*     *Possible Remedy*

### Frequent Starting and Stopping of Burner

(a) Thermostat adjustments incorrect

(a) Adjust thermostat in accordance with manufacturer's recommendations.

(b) Thermostat improperly wired.

(b) Check wiring and correct.

(c) Thermal element in thermostat loose.

(c) Screw thermal element in tight.

(d) Limit-control setting too low.

(d) Reset limit control to higher temperature.

(e) Nozzle too large for unit.

(e) Install nozzle of correct size.

(f) Dirty air filters.

(f) Replace filters.

### No Oil From Nozzle

(a) Plugged nozzle.

(a) Install new nozzle.

(b) Fuel supply exhausted.

(b) Refill supply tank.

(c) Leak in vacuum gauge.

(c) Repair as required.

(d) Leak in suction line.

(d) Tighten fittings or replace defective line.

(e) Leaky strainer gasket.

(e) Install new gasket.

(f) Pump not rotating.

(f) Inspect fuses, overload switch, wiring controls, and motor. Correct as necessary.

(g) Leak around pump shaft.

(g) Replace shaft seal.

## MAINTENANCE TIPS FOR YOUR CENTRAL SYSTEM
## WARM AIR HEATING EQUIPMENT

Obtain operation and maintenance information from the manufacturer of the equipment if possible. Different types of motors may need different service. In any case INSPECT THE EQUIPMENT AT LEAST ONCE A YEAR. Wipe off dust and if the motor has oil openings apply a few drops of nondetergent oil. Too much oil can be harmful because it collects dust and can result in overheating and interference with normal motor operation.

Inspect fan belts for wear and tension, replacing fan belts as may be required.

The air filter inspection is probably the most important and most frequent inspection requirement. The filter collects dust which restricts air flow, reducing the unit's overall efficiency. Most systems have disposable filters and these can be replaced. Clean the permanent washable type.

To improve overall heating efficiency, indoor climate control and comfort, and to reduce heating costs; install insulation in the attic, and walls; and in houses without basements, insulate the floors.

Give special attention to windows and doors. Those that open need weatherstripping. In most climates, storm windows and doors should be installed. *Thermopane* (insulating glass) windows installed in large window areas will cut heat loss in half through these building openings.

### STEAM-TRAP TROUBLE CHART

The following list will prove helpful in locating and correcting nearly all steam-trap troubles. Many of these are actually system troubles rather than trap troubles. Whenever a trap fails to operate and the reason is not readily apparent, the discharge from the trap should be observed. If the trap has been installed with a test outlet, the repair is simplified—otherwise it will be necessary to break the discharge connection.

### No Water or Steam Coming to Trap

(a) Plugged strainer ahead of or in trap.

(a) Clean or replace strainer.

| Symptom and Possible Cause | Possible Remedy |
|---|---|
| (b) Broken valve in line to trap. | (b) Replace valve. |
| (c) Pipe line or elbows plugged. | (c) Replace plugged unit. |

**Cold Trap—No Discharge**

| | |
|---|---|
| (a) Pressure too high. | (a) Reduce pressure or install traps with correct pressure rating. |
| (b) Wrong pressure rating originally specified. | (b) Replace with trap having correct pressure rating. |
| (c) Pressure raised without installing a pressure change assembly. | (c) Lower pressure or install pressure change assembly. |
| (d) Pressure regulating valve defective. | (d) Repair or replace valve. |
| (e) Pressure gauge on boiler reads too low. | (e) Recalibrate or replace pressure gauge. |
| (f) Orifice enlarged by normal wear. | (f) Repair or replace defective unit. |

**High Vacuum in Return Line**

| | |
|---|---|
| (a) Incorrect trap pressure differential. | (a) Increase pressure differential by installing correct pressure change assembly. |

**Trap Body Filled With Dirt**

| | |
|---|---|
| (a) No strainer in or ahead of trap. | (a) Install strainer. |

| *Symptom and Possible Cause* | *Possible Remedy* |
|---|---|
| (b) Excessive dirt in system. | (b) Purge system and remove source of dirt. |

## Bucket Vent Filled With Dirt Due to Oil or Other Water Condition

| | |
|---|---|
| (a) No strainer ahead of or in trap. | (a) Install strainer. |
| (b) Vent orifice too small. | (b) Enlarge orifice slightly. |
| (c) No scrubbing wire in bucket vent. | (c) Install scrubbing wire. |

## Hot Trap—No Discharge

| | |
|---|---|
| (a) No water coming to trap. | (a) Correct defect. |
| (b) Trap installed above a leaky bypass valve. | (b) Replace leaky valve or reposition trap. |
| (c) Broken or damaged siphon pipe in siphon-drained cylinder. | (c) Repair or replace as necessary. |
| (d) Vacuum in water-heater coils. | (d) Install 1/4″ check valve between steam-admission valve and coils. |

## Steam Loss

| | |
|---|---|
| (a) Valve not seating. | (a) Check valve seat for dirt or scale. Clear obstruction. Replace worn valve parts. |
| (b) Trap has lost prime. | (b) Close inlet valve for few minutes, then gradually open. If trap catches prime, it is OK. If not, replace. |

| *Symptom and Possible Cause* | *Possible Remedy* |
|---|---|
| **Continuous Discharge** | |
| (a) Trap too small. | (a) Install additional traps in parallel or replace with a larger unit. |
| (b) High-pressure trap used in a low-pressure system. | (b) Install correct pressure change assembly. |
| (c) Boiler foaming or priming. | (c) Install separator or correct feed-water conditions. |
| **Sluggish Heating** | |
| (a) One or more radiating units short-circuiting. | (a) Install a separate trap on each unit. |
| (b) Traps too small. | (b) Try next size larger trap. |
| (c) Air-handling capacity of trap is insufficient. | (c) Install a thermic bucket, or a larger-than-normal bucket vent. |

# CHAPTER 12

# Maintenance Management

The objective of a good maintenance program is to place into effect a plan whereby facilities and equipment are kept in good repair and in the proper operating condition at all times. In attaining this objective, there must be planning, scheduling, and supervision. Planning intent is to improve the work productivity, work control, and efficiency, all in an orderly manner.

## WORK STANDARDS

Many maintenance jobs are small, but after the time required to draw supplies, to travel to and from the job, and the time for any other purposes that might be necessary is all added, the accumulated time required to complete the job may be doubled. As an example:

*Fixing a screen door*

| | |
|---|---|
| Drawing Supplies | 6 minutes |
| Travel | 10 minutes |
| Actual Work | 15 minutes |
| Personal Time | 5 minutes |
| **Work Standard** | **36 minutes** |

In many instances, travel time can be reduced by establishing work zones and the time taken to draw supplies can be reduced by stocking certain repetitive-use items on a light transportation vehicle or hand cart.

A work standard for repetitive jobs can be established by placing all similar type work jobs together and, at the end of a given period of time, extracting the number of minutes used in performing the work and then dividing by the number of job orders.

*Example*:

Total time for replacing 40 ea. 48"
fluorescent light tubes ...............................320 minutes
Number of work orders with a 2-tube average
replacement requirement ............................. 20 work orders
Thus, each work order averages 16 minutes.

The 16-minute average time for replacing 2 each fluorescent light tubes can then be used as a work standard.

The average maintenance organization rarely has enough men to cope with all of the maintenance and alteration work requirements requested. However, since the prime mission is maintenance, this type work must be given priority if the facility is to function in an efficient manner. The supervisor is responsible for screening the work requests and establishing a priority of accomplishment. Through experience, he can usually compute the cost by establishing labor and material requirements. Some organizations have experienced estimators who establish the costs on other than repetitive work orders. These estimates can be used as an analysis for need and for scheduling the work.

It is assumed the average maintenance man, or the chief of the maintenance section, has a working knowledge of estimating. However, the following normal procedure is used in basic estimating for work performed by in-house forces:

1. First, the sketch or work request is examined.
2. The units are broken down by operation and trade, and all material and labor quantities are calculated.
3. These units are summarized to include supervision, transportation, and material taxes, if this is an item.
4. This then becomes the total job cost.
5. If the work is to be performed by contract, the contractors profit and overhead must be added.
   *Note*: An estimator should keep unit cost records from previously successfully estimated jobs for his use.

It is also assumed that the average maintenance leader has a knowledge of the common forms of mathematics. However, the following are ordinary formulas for calculating areas and volumes:

Area of a Square = length × breadth or height (Fig. 1)
Area of a Triangle = base × 1/2 of altitude (Fig. 3)

AREA = a x a

Fig. 1. Area of a square.

Area of a Parallelogram = base × altitude (Fig. 4)
Area of a Circle = square of diameter × .7854 (Fig. 5)
Circumference of a Circle = diameter × 3.1416 (Fig. 6)
Volume of a Cube = width × length × depth (Fig. 7)
Volume of a Cone or Pyramid = area of base × 1/3 of altitude
  (Fig. 8)

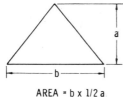

AREA = a x b

Fig. 2. Area of a rectangle.

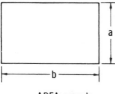

AREA = b x 1/2 a

Fig. 3. Area of a triangle.

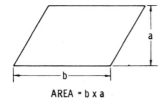

AREA = b x a

Fig. 4. Area of a parallelogram.

**Fig. 5.  Area of a circle.**

AREA = $D^2$ x .7854

**Fig. 6.  Circumference of a circle.**

CIRCUMFERENCE = D x 3.1416

**Fig. 7.  Volume of a rectangle.**

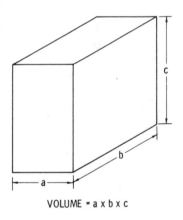

VOLUME = a x b x c

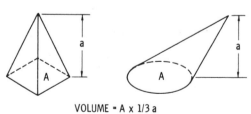

VOLUME = A x 1/3 a

**Fig. 8.  Volume of a pyramid or cone.**

323

## ESTIMATING

Estimators generally base their estimates on personal experience, but it is good practice to have available an estimator's handbook, several of which are commercially available.

If the job to be performed is other than the repetitive type (such as replacing a faucet washer), the estimator should provide a bill of materials for use by the person performing the work. If work sketches are not available for such jobs as partition and machinery installations and similar work, they should be prepared. Sketches and brief specifications enable workmen to know the exact requirements. Verbal instruction for work of this kind may result in costly errors.

## PLANNING AND SCHEDULING WORK

On work other than the repetitive type, a scheduling plan is necessary. Considering normal repetitive work, it can be assumed that 60% of the maintenance force will be available for assignment to work requirements such as major machine repair, painting rooms, replacing floors, installing partitions, etc. With approximately 60% of the crew available, this work can be scheduled for accomplishment.

Approximate man-hour labor requirements for major type work are known after work requests for this type work have been analyzed and estimated. Against this are balanced the available man hours (60% of the work crew), and the work can be scheduled and estimated completion dates stated on a work scheduling chart. After work of this type (scheduled work) has been completed, the supervisor can make an evaluation by comparing the estimated time against the actual time required for each particular job. This analysis will indicate the work efficiency and may suggest ways of work improvement.

Work orders are important records for purposes of preparing budgets, reviewing labor efficiency and material use, pin pointing repetitive trouble spots, and as a work-load review, as well as a review and back-up data for possible equipment purchases.

In conclusion, the most important factor in an efficient maintenance operation is good supervision. This determines the ultimate success or failure of the program. The best tools, materials,

A simple work-order form may be as follows:

Order No._____

Date:_____

Type of Work_____

Requested By_____

Assigned to_____

Start and Completion Time_____

Materials Used_____

and skilled craftsmen can be provided, but it remains the responsibility of supervision to see that these are used to the best advantage.

## ESTIMATING ROOF SURFACE AREAS

**Flat Roof**—Multiply length (a) by width (b). See Fig. 9.

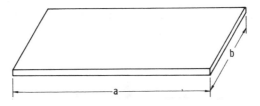

*Fig. 9. Figuring the area of a flat roof.*

*Example*—Length 50 ft., width 10 feet. $10 \times 50 = 500$ square feet or 5 squares of roof area.

**Gable Roof**—Multiply length (b) by width (a), and add 1/4 of the total. See Fig. 10.

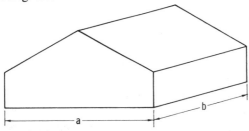

*Fig. 10. Figuring the area of a gable roof.*

*Example*—Length 50 feet, width 10 feet. 10 × 50 = 500 + (1/4 of 500) = 500 + 125 = 625 square feet or 6-1/4 squares of roof area.

**Arch Roof**—Multiply length (b) by width (a) and add 1/2 of the total. See Fig. 11.

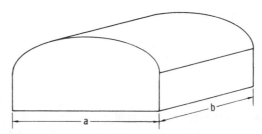

*Fig. 11. Figuring the area of an arch roof.*

*Example*—Width 10 feet, length 50 feet. 10 × 50 = 500 + (1/2 of 500) = 250 + 500 = 750 square feet or 7-1/2 squares of roof surface.

**Gambrel Roof**—Multiply length (b) by width (a) and add 1/3 of the total. See Fig. 12.

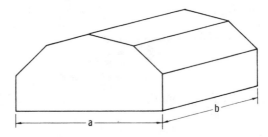

*Fig. 12. Figuring the area of a gambrel roof.*

*Example*—Width 10 feet, length 50 feet. 10 × 50 = 500 + (1/3 of 500) = 500 + 167 = 667 square feet or 6.67 squares of surface area.

CHAPTER 13

# Custodial Practices

Clean surroundings in a building have an effect on both health and morale of the occupants. In addition, proper cleaning of floors and wall surfaces protects them from damage and maintains the intended original surface.

## FLOORS

Floors are cleaned more than any other portion of an average structure. The most common method of cleaning is by the use of a broom or vacuum cleaner. The type of cleaning equipment will vary depending on the type of floor covering and the use of the facility. The following are some recommended sweeping tools for various types of floors:

| Type of Floor | Sweeping Tool |
| --- | --- |
| Rough concrete | Fiber floor brush |
| Treated or painted smooth concrete | Untreated sweeping mop |
| Oxychloride cement | Untreated sweeping mop |
| Linoleum, asphalt, vinyl, terrazzo, mosaic, quarry, and rubber tile, and smooth, sealed, and waxed floor | Untreated sweeping mop |

Smooth, sealed, or painted wood floor      Treated sweeping mop

Unpainted or unsealed wood floor      Fiber floor brush

It is essential that all dust and dirt gathered by sweeping mops or brooms be picked up in a dust pan and emptied into a container provided for that purpose.

## CLEANING FREQUENCY

The frequency of sweeping is determined by the use of the facility and local conditions such as weather, condition and types of exterior walks, soil, and general traffic.

Administration offices, cafeterias, hospitals, laboratories, and even some shops where fine precision work is being done, should be swept and dust mopped at least once daily. Wet mopping should be accomplished at least every week. Buffing should be scheduled for administration and general office sections, cafeterias, and hospitals at least once a week.

Warehouses should be swept on a twice-a-month cycle, depending on specific local conditions.

Toilet floors should be swept and wet mopped daily, or as often as use and traffic requires. Toilet water closets, urinals, wash basins, sinks, and drinking fountains should be cleaned daily.

Desks, office chairs, bookcases, and files should be dusted daily, venetian blinds once a month, and light fixtures every month or once every two months. Walls, ceilings, and exposed piping, three times a year. Radiators, once a month. Air-conditioning grilles at least every month.

Carpets in offices should be vacuumed every day or twice a week.

Partitions and woodwork (excepting toilets) should be washed yearly. Spot clean weekly, or as required.

Stairways and entrances should be swept and mopped daily.

Sand urns, ash trays, and wastebaskets should be cleaned daily.

Trash should be removed as required.

## CLEANING COMPOUNDS

Detergents are in common use, and nearly any brand can be used effectively on asphalt and rubber tile, linoleum, and on painted and lacquered surfaces.

Scouring powder may be used on ceramic and quarry tile, but only with extreme care; it should not generally be used on linoleum, unglazed floor tile, marble, wood floors, and on painted areas. The abrasives in this type of cleaner may scratch the surfaces to which it is applied.

Trisodium phosphate, a strong cleaning agent, may, if diluted with water, be used for removing grease and for washing soiled painted surfaces. Its use should be limited, and a complete rinsing with water is recommended after the washing.

In using cleaning agents, it is recommended that no stronger solution be used than is absolutely necessary, and that the surfaces be rinsed with clear water after use of the cleaner. Also, it is recommended that scouring powders and steel wool be used with care to prevent damage to the surfaces.

### MOPPING

Soap and water generally soften dirt that can not be removed with a broom. A mop is used to spread the detergent-water solution, to rub loose any sticking dirt, and later to remove this solution and the rinse water.

Care must be used with the application of water on floors, as excessive water and strong mopping solutions may cause possible damage. It is best to mop, rinse, and dry small areas at one time. Applying water and mopping solutions to large areas results in the water being on the floor for longer periods of time than is necessary, causing possible floor damage.

Do not splash furniture, equipment, baseboards, and radiators. Mop all hard-to-reach places by hand.

Strong detergents and strong solutions of trisodium phosphate may soften wood fibers, so care must be used in mopping untreated or unsealed wood floors. Use approximately 1/6 cup of synthetic detergent to a gallon of water and wet mop only if the floor is very dirty. Use as little water as possible when cleaning floors of this type.

A similar procedure should be used for cleaning sealed or painted wood surfaces which have not been waxed.

When cleaning waxed wood, linoleum, or asphalt tile floors, use clear water and about 1/10 cup of synthetic detergent. Floors should be mopped only if they cannot be dry cleaned.

When mopping terrazzo, quarry tile, flagstone, mosaic tile, slate, marble, or unpainted concrete, use a weak solution of all-purpose synthetic detergent mixed with water. The washing solution should be changed often. Strong washing solutions should be avoided as they may result in pitting and rapid floor deterioration. Floor *scrubbing* should be performed only when dirt cannot be removed by mopping. Scrubbing should be kept to a minimum.

## STAIN REMOVAL

There are many ways of removing stains, but among the most common are the following:

*Water Stains*—For removing water stains from urinals, toilet bowls, and washbowls, a daily cleaning with an all-purpose synthetic detergent will suffice. Bad stains can be removed with a weak solution of toilet-bowl cleaner or with a grit-type soap. Do not use toilet-bowl cleaner more often than necessary, as it may eventually damage the porcelain glaze on the fixture. Muriatic and similar type acids quickly damage the porcelain fixture glaze and should not be used at all.

*Ink Spots*—On wood, asphalt tile, concrete, marble, and terrazzo, apply a solution of 1 part oxalic acid crystals to 9 parts of warm water. Allow the solution to dry and then mop with clear water.

On vinyl, wash the surface with a synthetic detergent, rinse, and then dry. If this does not remove all of the ink, rub the spot with a cloth dampened with regular ammonia.

*Grease or Oil*—For asphalt tile and linoleum, wash the surface with an all-purpose synthetic detergent solution and rinse with clear water. For wood and vinyl surfaces, carefully pour some kerosene on the spot, allow it to soak for a short time, and then wipe with a clean cloth; then wash with an all-purpose synthetic detergent solution and rinse with clear water. On concrete, apply alcohol to the spot, rub, and then wipe dry with a clean cloth.

*Iodine or Mercurochrome*—For removing iodine or mercurochrome from wood, linoleum, and concrete, apply alcohol and wipe with a clean cloth. On vinyl tile, wash with an all-purpose synthetic detergent, rinse, and wipe dry. If the stain is not completely removed, apply scrubbing powder and rub with a cloth and warm water.

*Blood*—On linoleum, asphalt tile, vinyl tile, and wood, rub the blood-stained spot with a cloth dampened in clear, cold water. If the stain is difficult to remove, dampen the cloth with ammonia and rub the spot. On marble, concrete, and terrazzo, rub the spot with a cloth dampened in cold water. If this does not remove the stain, bleach with peroxide.

*Heel and Sole Marks*—For wood, linoleum, vinyl, and asphalt tile, wash the heel and sole marks with an all-purpose synthetic detergent solution. If the marks persist, rub them with No. 0 steel wool and then rinse with clear water.

For concrete, marble, and terrazzo, wash the marks with a good detergent and rinse with clear water.

*Rust*—For wood, linoleum, asphalt, and vinyl, wash the rust spot with an all-purpose detergent and rub with No. 0 steel wool as may be required. On marble, unglazed tile, and concrete, an application of any good commercial rust remover sold at your local hardware store should remove the stain.

Rust can be removed from fabrics by applying very hot water; place fresh lemon juice on the area and then rinse after 4 minutes. Repeat the process as may be required.

*Chewing Gum*—On wood, marble, terrazzo, concrete, and linoleum, remove as much of the gum as possible with a putty knife, and then apply denatured alcohol, rubbing with a clean cloth. On asphalt and vinyl tile, remove the gum with a putty knife, and then rub carefully with No. 0 steel wool dipped in an all-purpose synthetic detergent solution.

On fabrics, scrape off all the chewing gum possible with a putty knife, and then apply a commercial cleaner.

*Tar*—On concrete, wood, linoleum, and vinyl tile, remove as much of the tar as possible with a putty knife, carefully apply kerosene and then wash with an all-purpose synthetic detergent solution. On asphalt tile, remove the tar with a putty knife and then wash with an all-purpose synthetic detergent solution. *Do*

331

*not use kerosene on asphalt tile.* On marble and terrazzo, remove as much of the tar as possible with a putty knife, soak with denatured alcohol, and then cover with Fuller's earth and allow to dry, wiping the surface clean after about an hour.

*Paint*—For linoleum, marble, terrazzo, and vinyl tile, rub the paint spot with No. 0 steel wool dipped in turpentine or kerosene; then wash with an all-purpose detergent solution and rinse with clear water.

For asphalt tile, rub with steel wool and an all-purpose synthetic detergent, and rinse with clean water.

On concrete and wood floors, wash with a solution of 1/2 lb. of trisodium phosphate to 1 gallon of warm water. Rinse with clean water. *Use rubber gloves to protect the hands.* Use water sparingly on wood floors and wipe dry as quickly as possible.

Fresh paint can generally be removed from fabrics by washing with water and soap. When paints are set and hardened, soaking may be required with turpentine or household ammonia. Shellac stains may be removed by sponging with alcohol.

## DRY CLEANING FLOORS

Some building custodial personnel use cylinder-type electrically operated machines for cleaning floors. These are equipped with a steel-wool or brush cylinder and a vacuum attachment with a wax bar riding the top of the steel-wool or brush cylinder. If the floors are sealed and in good condition, this machine can clean, wax, and buff in one operation, thereby reducing maintenance time. It has interchangeable accessories, complete with a built-in vacuum system. It can be used for polishing, dry buffing, scrubbing, and also as a combination sweeping and polishing device. It is especially adaptable for use in corridors, offices, and large areas.

Small disc-type machines with steel-wool pads are available for use in small areas.

Dry cleaning is recommended instead of wet mopping, as the floors are less likely to be damaged by the use of this process.

Dry cleaning is not recommended for pine or fir wood floors which have not been sealed, or for splintered wood floor areas.

## FLOOR WAXES

On wood floors which are not sealed or varnished, the use of bar wax is recommended. On wood floors which are well sealed or varnished, the use of bar wax is preferred, but water-emulsion wax may be used. For asphalt tile, cork, vinyl, rubber, linoleum, and oxychloride-type floors, bar wax, water-emulsion, and resin-emulsion waxes may be used.

It is not recommended that the wax be applied closer than 6 or 8 inches to walls and partitions, as the buffing machine will spread the wax to these little-used areas. This will help prevent excessive wax build-up on the floors.

Heavy-traffic areas should be waxed as often as necessary to protect the floors and/or floor coverings. Wax must be applied in a thin, even coat, and the floors buffed to a uniform sheen in all areas, including under desks and other furniture.

Wax must be periodically removed from the floors to prevent difficulty in cleaning and to maintain a good appearance. There are several good commercial wax-remover solutions on the market. If none is available, synthetic detergent in solution with ammonia and water may be used, but with extreme care, as it may damage the flooring. After application, remove the wax and solution as quickly as possible and mop or rinse with clear water, using water sparingly. Wax-removal solutions are primarily made for synthetic-type floor coverings, such as asphalt, rubber, linoleum, and vinyl.

## WASHING WALLS

Before the start of the cleaning operations, move all furniture away from the walls and remove all pictures and other wall furnishings which may interfere with the washing operation. Dust all areas to be washed. Use a 14-quart pail filled 3/4 full and mix into it about 1/2 cup of all-purpose detergent. Place buckets on paper to prevent possible floor damage and place sponges in the buckets. When using sponges, do not wring them dry, but merely squeeze until they do not drip. Continuous and heavy forceful wringing can easily ruin a sponge.

Start washing in a corner, and continue until an area about 5-feet wide and halfway to the ceiling is cleaned, rinsing the sponge often and working up with straight, even strokes. When a width of about 10 feet has been washed, start work on the upper half in the same corner where the lower half work was started, again using straight even strokes, squeezing the sponge as required after placing it in the bucket, but only so that it does not drip. Do not allow water to run on the lower washed section of the wall. Continue this operation, working from the lower to the upper section until the entire wall has been washed and is free of streaks and water marks. Replace all furnishings.

*Caution*: *Extreme care must be used when working with ladders or planks, as careless handling may result in personal injury or damage to furnishings and/or the building walls and glass.*

Good wall-washing practices reduce painting frequency requirements and, at the same time, provide well–appearing surfaces. Painted surfaces may be washed many times before repainting is necessary if washing is properly done. Wall-washing materials include an all-purpose synthetic detergent, sponges or clean cloths for washing and rinsing, and also a clean cloth for drying. Two 14-quart buckets are needed per man for washing and rinsing. Soap and water solutions and rinse water must be changed frequently to assure clean surfaces.

## Wall Finishes

*Glass and Glazed Tile Walls*—A good neutral cleaning solution mixed with water works well on these surfaces. Strong cleaners and abrasives are not recommended as they may damage surfaces and mortar joints.

*Accoustical Walls*—Walls of this type are usually cleaned with a vacuum cleaner, although the surface may be washed if painted, but only after cleaning the holes or depression in the material with a vacuum cleaner. Only small areas should be cleaned at a time and quickly wiped dry. Start on the ceiling and work down. Material of this type is easily damaged and scuffed. Wash only if absolutely necessary and then with extreme care. Do not wash unpainted accoustical tile.

*Marble, Granite, or any Natural Stone Wainscot*—Wash with mild soap or detergent and water solution, rinse with clear water, and dry with a clean soft cloth.

*Unpainted Plaster Walls and Walls Painted With Water Paints*—A vacuum cleaner should be used on unpainted plaster walls. Water should not be used as the porous surface will quickly absorb the water and damage the plaster.

Walls painted with whitewash should not be washed with water, although some of the present day latex water-type paints can be washed. A small section of the wall should be tested prior to washing. Wallpaper cleaners are recommended for casein painted surfaces.

*Oil Painted and Enameled Walls*—Wall surfaces of this type can be washed with almost any all-purpose mild detergent and water without damage, if care is used in the washing procedure.

On gypsum walls (plasterboard), clean any scarred or scuffed surfaces with a wallpaper cleaner as water seeping into the open sections, including untaped seams, may result in discoloration and possible damage. Taped and unscarred surfaces may be washed if care is used.

### Washing Woodwork

Prior to washing woodwork, dust must be removed, particularly above doors and on window sills. Remove items such as shades, drapes, or other furnishings that could possibly interfere with the work. Start at the bottom and wash upward, using straight, even strokes. When a small area has been washed, go over the surface with a rinsing sponge after it has been squeezed so it will not drip. Pick up as much water as possible and then wipe the area dry with a clean cloth. After the lower portion of the work has been completed, start on the upper section, using the same procedure. The entire surface should be clean, without any streaks or spots.

### DUSTING

Although it is common practice to use a feather duster, this procedure is not recommended as it only spreads the dust. Dusting should be done with a treated cloth or a yarn duster, and the dust from these shaken into a dust box or container. When dust cloths or yarn dusters become too dirty for further use, they can be washed in a solution consisting of a tablespoon of trisodium phosphate added to a gallon of warm water. Hang the dusters to dry in a location where there is good air circulation.

A vacuum cleaner, preferably of the industrial type, complete with tube and soft bristle brush attachments can be used for dusting high places and walls, as well as locations difficult to reach. High places should be dusted before low and easy-to-reach locations, and before floors are swept.

Typewriters, adding machines, and similar type equipment should not be cleaned by custodial personnel, particularly if the equipment is not protected with covering device.

Tables and desks should be cleaned with long, even strokes, using treated dust cloths or hand dusters and holding the cloth lightly so it absorbs the dust. Do not disturb papers left on desks, but do lift letter trays, books, and similar items to dust under them. Do not leave dust streaks, and *do* dust chair legs and rungs.

## WASHING WINDOWS

It is best to use clear water to wash windows. If the windows are exceptionally dirty, some detergent with a very small amount of kerosene may be added to a gallon of water and this solution used. This mixture cuts dirt, grease, oil film, and leaves no deposit or streaks on the glass. A few teaspoons of trisodium phosphate in a gallon of water also makes a good cleaning solution. Adjacent painted surfaces should be washed clean if the cleaning solution splatters on it. Good commercial window-cleaning solutions are available, but some are expensive. Ammonia should not be used, as it dries out and loosens putty.

Windows should not require washing more than two or three times a year under normal conditions. However, frequency requirements depend primarily on use of the building, season of the year, and location of the facility with respect to smoke, dust, and traffic conditions.

Windows should be cleaned at a time that will result in a minimum of interruption to the occupants. Care must be used not to spill water on the sash, window sills, and furnishings. The use of squeegees for large windows reduces labor costs and surface drying time. Small windows may be washed with a clean cloth dampened in the washing solution and dried with a chamois.

## TOILET ROOMS

Under normal conditions, a daily cleaning of toilet rooms is recommended. Specific conditions will determine the cleaning frequency. In all instances, a high standard of cleanliness should be maintained.

### Lavatories and Drinking Fountains

Make a mild cleaning solution by placing an all-purpose detergent in a half-gallon of warm water. (Water requirement depends on the size of the toilet room). Remove bars of soap from the lavatory and remove all paper, chewing gum, and other litter from the interior of the fixtures. Dampen a cloth in the solution and clean the fixtures, rubbing hard enough to remove any deposits. Remove stains with a mild grit cleaner. Wipe the fixtures clean with a dry cloth. Avoid the use of strong cleaners or acids.

### Toilets and Urinals

Improper and/or infrequent flushing of urinals causes crystallization of urine salts above the trap water level, resulting in offensive odors. Flushing cannot be over emphasized, as the flushing action after each use prevents urine salt crystal deposits.

Fixtures and toilet seats should be washed clean with a mild solution of all-purpose detergent and warm water. For removing stains (including those from hard water), a tablespoon of trisodium phosphate mixed in a gallon of warm water can be used. Care should be used with this solution as continuous use may damage porcelain or vitreous-china fixture coatings. Strong cleaners and acids damage fixture surfaces and should not be used over a prolonged period of time.

For cleaning the interior surfaces of urinals and toilet bowls, use a toilet brush, working as far as possible into the toilet trap. The inside rim of the toilet bowl should be thoroughly cleaned. Inspection can be made with a hand mirror. Flush the fixtures after cleaning.

### Floors, Partitions, and Walls

It is essential that toilet-room floors be mopped at least once a day to prevent odors. An all-purpose detergent mixed with water

337

can be used. Scrub the urinal areas as may be required, as most floor substances absorb urine and create offensive odors.

Wall and partition dirt accumulations should be removed daily, and all partitions and walls washed as often as required.

### Dispensers

Every effort should be made to keep toilet-room dispensers filled at all times. Material substitutions, particularly in the case of toilet paper, can result in expensive repairs. Paper towels have been known to clog toilets, resulting in water overflow, inconvenience, and costly repairs.

## RUG CLEANING

Commercial or specially trained personnel are recommended for particularly dirty or soiled rugs where shampooing may be required. Normal cleaning can be accomplished by the regular custodial personnel using a vacuum cleaner. Regular personnel can also lightly shampoo rugs and carpeting with some of the common rug detergents and special shampooing machines which are readily available on loan from many hardware store and carpet dealers.

Although household-type vacuum cleaners work well for cleaning carpeting, the industrial type with a blower vacuum has greater air intake and is more efficient for overall use. Household-type machines have small air intakes, but work well where areas to be cleaned are comparatively small.

## PERSONNEL REQUIREMENTS

Personnel requirements will vary depending on the building location, its use, and its category. In establishing frequency schedules, buildings should first be grouped by categories or types, such as:

Production (garages and shops)
Administration and Personnel (offices, laboratories, cafeterias, and drafting rooms)
Storage areas (storerooms, warehouses, etc.)

A listing is then made of the required frequencies, such as *Daily, Twice Weekly, Weekly, Twice Monthly,* etc.

Type of work and possible work difficulties and estimated cleaning time are listed as follows:

It is a good policy to place a metal card holder in the toilet room containing a lined paper card so that responsible personnel can initial and indicate the time and date when the fixtures were cleaned.

List all work items

*Floors*

| Room x | Obstructions | Move Furniture | Cleaning Time | Frequency |
|---|---|---|---|---|
| 1200 sq. ft. | | | | |
| Sweep | 2 columns | chairs | 24 minutes | D (Daily) |
| Vacuum | 2 columns | chairs | 23 min. | 2 Wk (Twice weekly) |
| Wet Mop | 2 columns | chairs | 43 min. | 1 Wk |
| Buff | 2 columns | chairs | 32 min. | 1 Wk |

Continue with the type of work and the possible items that would effect scheduled work.

*Cleaning Toilet Rooms*
(Men's in area x—200 sq. ft.)

| Fixture Type | Total No. | Cleaning Time | Frequency |
|---|---|---|---|
| Urinal | 2 | 5 min. ea. | D |

The general chart schedule can be continued until all requirements and frequencies have been listed. These are to include glass cleaning, dusting, rug cleaning, stairway and entrance cleaning, emptying sand urns, trash containers, and room wastebaskets, cleaning woodwork, venetian blinds, light fixtures (incandescent or fluorescent), radiators, air-conditioning grilles, bookcases, files, desks, chairs, and all allied cleaning requirements. These requirements should be scheduled by the number of estimated minutes and/or hours each work item requires.

After the work items have been determined, and the estimated work time and work cycle (daily, twice-weekly, once-weekly, monthly, etc.) scheduled, the personnel requirements may be approximated.

The work month can be calculated as being 173 hours and the work year as being 2,080 hours on a 40-hour, 52-week basis.

The following time-table lists some cleaning approximations. These may vary somewhat, depending on furniture layout, obstacles, etc.

| | |
|---|---|
| Sweep (general rooms) | 45 sq. ft. per minute |
| Sweep (stairways) (50 steps and landing) | 15 minutes |
| Sweep (unobstructed halls and corridors) | 80 sq. ft. per minute |
| Damp Mop | 80 sq. ft. per minute |
| Wet Mop and Rinse | 30 sq. ft. per minute |
| Hand Scrub | 175 sq. ft. per hour |
| Machine Scrub (16-inch floor machine, unobstructed area) | 900 sq. ft. per hour |
| Deck Scrub Brush (with long handle) | 500-600 sq. ft. per hour |
| Wax | 1600-1800 sq. ft. per hour |
| Machine Polish (16-inch wax machine in unobstructed area) | 50 sq. ft. per minute |
| General Office Dusting | 66 sq. ft. per minute |
| Dusting Classrooms (average 40 desks with 1 table, 2 chairs, 6 window sills) | 8 minutes per classroom |
| Washrooms (cleaning bowls, floor, lavatories, etc.) | 400 sq. ft. per hour |
| General Office Cleaning (women) | 1000 sq. ft. per hour |
| Washing Marble Walls (sponge and squeegee) | 550 sq. ft. per man hour |

| | |
|---|---|
| Window Washing (per man) | 70 small or average size windows per day (8 hrs.) |
| Washing Painted Walls (varies depending on soil) | 350 sq. ft. per hour |

## SAFETY

A good safety program with emphasis on good safety practices must be continuously maintained. All ladders, safety belts (if used), planks (if used), and electrical devices (vacuum cleaners, buffing machines, etc.) must be maintained in good safe condition. Personnel must be instructed to keep mopping pails, brooms, and other cleaning items out of the way of working personnel, particularly in traffic lanes during working hours. Good housekeeping is essential to maintain safe working conditions and to prevent fires. Combustible material must not be allowed to accumulate.

# Glossary

**Acoustical Tile**—A special tile for walls and ceilings made of wood, mineral, vegetable fibers, metal or cork. Its purpose is to control sound volume, while providing cover.

**Air Duct**—Pipes that take warm and cold air to rooms and back to the furnace or air conditioning system.

**Ampere**—The rate flow of electricity through electric wires.

**Apron**—Generally a paved area, such as a garage entrance or the juncture of a driveway with the street.

**Backfill**—Earth or gravel replaced in the space around a building wall or any excavation after pipe or foundations are in place.

**Balustrade**—A row of ballusters topped by a rail, edging a balcony or a staircase.

**Baseboard**—A board along the floor against walls and partitions in a building to hide gaps.

**Batt**—Insulation in the form of a blanket, rather than loose filling.

**Batten**—Small thin strips covering joints between wider boards on exterior building surfaces.

**Beam**—One of the principal horizontal steel or wood members of a building.

**Bearing Wall**—A wall that supports a floor or roof of a building.

**Bib or Bibcock**—A water faucet to which a hose may be attached. Also called a sill cock or a hose bib.

**Bleeding**—Seeping of gum or resin from lumber. This term is also used in referring to the process of drawing air from a water pipe.

**Brace**—A piece of wood, steel or other material used to form a triangle and stiffen some part of a structure.

**Braced Framing**—A construction technique using posts and cross bracing for greater rigidity.

**Brick Veneer**—Brick used for the outer surface of a framed wall.

**Bridging**—Small metal or wood pieces placed diagonally between floor joists.

**Built-up Roof**—Roofing material applied in sealed, waterproof layers. Generally applied on nearly flat and/or roofs with a slight slope.

**Butt Joint**—The joining point of two pieces of molding or wood.

**BX Cable**—Electrical cable with a rubber coating and a flexible steel outer covering.

**Cantilever**—A projecting joist or beam, without support at one end, used to support an extension of a structure.

**Carriage**—That member which supports the steps or treads of a stair.

**Casement**—A window sash that opens on hinges placed along the side (vertical edge).

**Casing**—Window and door framing.

**Cavity Wall**—A hollow wall formed by linked masonry walls providing an air space between the walls.

**Chimney Cap**—Concrete capping around the top of chimney bricks for purposes of protecting the masonry from the elements.

**Chair Rail**—Wooden molding placed on a wall around a room at the level of a chair back.

**Chase**—A groove in a masonry wall or through a floor to accommodate ducts or pipe.

**Circuit Breaker**—A safety device used in electrical work which opens an electrical circuit automatically when it becomes overloaded.

**Cistern**—A tank used to catch and store rain water.

**Collar Beam**—A horizontal beam fastened above the lower ends of rafters to add rigidity.

**Coping**—Brick or tile used to cap or cover the top of a masonry wall.

**Corbel**—A horizontal projection from a wall, forming a ledge or supporting a structure above the ledge.

**Corner Bead**—A wood or metal strip protecting the outside corners of plastered walls.

**Course**—A horizontal row of cinder blocks, bricks, or other masonry materials.

**Cove Lighting**—Light sources concealed behind a cornice or horizontal recess which directs the light upon a reflecting ceiling.

**Crawl Space**—An unfinished space beneath the first floor of a building which has no basement and the ground, used for visual inspection and access to ducts and pipe.

**Door Buck**—The rough frame of a door.

**Dormer**—The projecting frame of a recess in a roof slope.

**Double Glazing—Thermopane windows**—An insulating window pane formed of two thicknesses of glass with a sealed air space between them.

**Double-Hung Windows**—Windows with upper and lower sash, each supported by cords and weights.

**Downspout**—(Conductor pipe) A pipe used to carry rainwater down from a roof or gutters.

**Drip**—Projecting part of a cornice, which sheds rain water.

**Dry Wall**—A wall surface of material other than plaster, usually plasterboard.

**Eaves**—Roof extension beyond house walls.

**Efflorescence**—The white powder that forms on the surface of bricks.

**Effluent**—Treated sewage from a sewage treatment plant or from a septic tank.

**Fill-Type Insulation**—Loose insulating material applied by hand or mechanically blown into wall spaces.

**Flashing**—Metal used around angles or junctions on exterior walls, chimneys, or roofs to prevent leaks.

**Floor Joists**—Framing pieces resting on outer foundation walls and interior beams or girders.

**Flue**—A chimney passageway for conveying gases, smoke, and fumes to the outside air.

**Footing**—The concrete base on which a foundation sits.

**Foundation**—Lower wall parts on which a structure is built. Masonry and concrete foundation walls are usually below ground level.

**Framing**—Joists, studs, rafters, and beams—the rough lumber of a building.

**Furring**—This wood or metal applied to a wall or floor to level the surface for lathing, boarding, or plastering.

**Fuse**—A safety device in an electrical panel box which breaks (opens) an electrical circuit when it becomes overloaded.

**Gable**—The triangular part of a wall under the inverted "v" of the roof line.

**Gambrel Room**—A roof with two pitches. The roof is flatter toward the ridge and is steeper on its lower slope.

**Girder**—A main structural member carrying the weight of a floor or partition.

**Glazing**—Fitting glass into windows or doors.

**Grade Line**—Generally the point at which the ground rests against the foundation wall.

**Green Lumber**—Lumber which has not properly dried and which tends to warp or "bleed" resin.

**Gusset**—A bracket or brace used to strengthen a structure.

**Gutter or Eave Trough**—A shallow channel of metal set below and along the eaves of a house to catch and carry off rainwater from the roof. In barns, the depression in a floor.

**Headers**—A stone or brick extending over the thickness of a wall. A beam placed perpendicular to joists and to which joists are nailed in framing.

**Heel**—The end of the rafter that rests on the wall plate.

**Hip**—The external angle formed by the juncture of two slopes of a roof.

**Herring-Bone Work**—Bricks, tile or other materials laid slanting in opposite directions.

**I-Beam**—A steel beam with a cross section resembling the letter "I".

**Insulating Board or Fiberboard**—A low-density board made of wood, sugar cane, cornstalks, or similar materials, usually formed by a felting process, dried and pressed to a specified thickness.

**Insulation**—Any material high in resistance to heat transmission that, when placed in the walls, ceilings, or floors of a structure, will reduce the rate of heat flow.

**Jalousies**—Windows with movable, horizontal glass slats angled to admit ventilation and keep out rain. The term is also used for outside shutters of wood constructed this way.

**Jamb**—The side post or lining of a doorway, window or other opening.

**Joist**—One of a series of parallel beams used to support floor, roof, and ceiling loads, and supported in turn by larger beams, girders, or bearing walls. Members supporting roofs having slopes over 3 in 12 are called rafters.

**Keystone**—The stone place in the center of a span or the top of an arch.

**Kiln-dried**—Artificial drying of lumber, generally superior to most lumber that is air dried.

**King-post**—The middle post of a truss.

**Lag screws**—Large, heavy screws, used where great strength is necessary, as in heavy framing or when attaching ironwork to wood or concrete.

**Lally column**—A steel pipe or tube sometimes filled with concrete, used to support girders or other floor beams.

**Landing**—A platform between flights of stairs or at the termination of a flight of stairs.

**Lath**—A building material of wood, metal, gypsum, or insulating board on which plaster is spread.

**Leaching bed**—Tiles in the trenches carrying treated wastes from septic systems.

**Lean-to**—A small building whose rafters pitch, or lean, against another building, or against a wall.

**Light**—Space in a window sash for a single pane of glass. Also, a pane of glass.

**Ledger**—A piece of wood which is attached to a beam to support joists.

**Lintel**—The horizontal structural member which spans a door, window or other opening to carry the weight of walls above.

**Load-bearing wall**—A strong wall which is capable of supporting weight.

**Louvres**—The inclined slats spaced at intervals to admit a free air current and at the same time exclude the rain.

**Masonry**—Stone, brick, concrete, hollow-tile, concrete block, gypsum-block, or other similar building units or materials or a combination of the same, bonded together with mortar to form a wall, pier, buttress, or similar mass.

**Miter**—The joining of two pieces at an angle that bisects the angle of junction.

**Molding**—A strip or piece of decorative material having a plane or curved narrow surface prepared for ornamental application. These strips are generally used to hide gaps at wall junctures.

**Monitor**—That portion of a building extending above the main roof for the purpose of ventilating or lighting the interior.

**Moisture barrier**—Treated paper or metal that retards or bars water vapor, used to keep moisture from passing into floors or walls.

**Monolithic concrete**—Poured (solid) concrete.

**Mullion**—Framing of slender material which divides the lights of panes of windows.

**Newel**—The upright post or the upright formed by the inner or smaller ends of steps about a winding staircase. In a straight staircase, the principal post at the foot or the secondary post at a landing.

**Nosing**—The rounded and projecting edge of the treads of a stair, or the edge of a landing.

**Offset**—A ledge occuring at a change in thickness or width of a wall.

**On Center**—The measuring of spacing for studs, rafters, joists, and the like in a building from center of one member to the center of the next member.

**Overhang**—A projection of an upper part (as a roof, upper story) of a building extending beyond the lower part.

**Parging**—A rough coat of mortar applied over a masonry wall as protection or finish.

**Pent House**—A roof structure covering stairways, elevator shafts, etc.

**Pilaster**—A projection of the foundation wall used to support a floor girder or to stiffen the wall.

**Piling**—Structural members sunk into the ground and used to support vertical loads.

**Pitch**—The angle slope of a roof expressed in inches of rise per foot of run.

**Plates**—Horizontal wood members which provide bearing and anchorage for wall, floor ceiling and roof framing.

**Plenum**—A chamber which can serve as a distribution area for heating and cooling systems. Usually that chamber directly above the furnace.

**Pointing**—Treatment of joints in masonry by filling with motar to protect against weather or to improve appearances.

**Prefabrication**—Components such as walls, trusses, or doors are constructed before delivery to the building site.

**Purlin**—A small longitudinal beam resting on trusses and supporting the roof.

**Rabbet**—A groove cut in a board to receive another board.

**Rafter**—A series of roof framing members. Members supporting roofs having slopes 3 in 12 or less are defined as roof joists.

**Ramp**—An inclined walk or roadway.

**Reinforced concrete**—Concrete strengthened with metal bars or wire mesh.

**Ridge pole**—A longitudinal plank to which the ridge rafters of a roof are attached.

**Riser**—Each of the vertical boards closing the spaces between the treads of stairways.

**Roof Covering**—Any type of material put on a roof to make it watertight.

**Roof Sheathing**—Boards or sheets (sheets are usually plywood) which are nailed to the top edges of trusses or rafters to tie the roof together and to support the roof material.

**Rubble Masonry**—Masonry laid without respect to uniformity of courses and joints.

**Sash**—A single frame containing one or more lights of glass.

**Section**—A drawing showing the parts of a building as they appear if the building were cut through vertically.

**Scotia**—A concave molding.

**Scuttle hole**—A small opening either to the attic, roof, or to the crawl space.

**Shake**—A hand or machine split wooden shingle, usually edge grained.

**Shim**—Thin tapered pieces of wood used for leveling or tightening a stair or other building element.

**Siding**—The finish covering of the outside wall of a frame building.

**Sill plate**—The lowest member of the building framing resting on top of the foundation wall. Sometimes called the mud sill.

**Skylight**—A glass opening in the roof.

**Sleeper**—Timber laid on the ground to receive joists, or strips of wood laid over concrete floor to which the finished wood floor is nailed or glued.

**Soffit**—The underside of the members of a building such as stairways, cornices, beams, and arches, relatively minor in area as compared with ceilings.

**Soil stack**—A general term for the vertical main of a system of soil, waste, or vent piping.

**Span**—The distance between the supports of a beam, girder, arch, truss, etc.

**Stud**—One of a series of vertical wood or metal structural members placed as supporting elements in walls and partitions. (Plural: studs or studding).

**Subfloor**—Boards or sheet material laid on joists over which a finished floor is to be laid. Also often termed floor sheathing.

**Sump**—A pit in the basement in which water collects to be pumped out by a pump.

**Swale**—A wide shallow depression in the ground to form a channel for storm water drainage.

**Terra cotta**—Glazed clayware used in the facing of buildings.

**Terrazzo**—A highly polished flooring made of cement and marble chips.

**Tie**—A timber, rod, chain, etc., holding two or more structural members together.

**Tile field**—Open-joint drain tile laid to distribute septic tank effluent over a preplanned area or to provide subsoil drainage in wet areas.

**Toenail**—Driving nails at an angle into corners or other joints.

**Transom**—The sash over a doorway.

**Trap**—A bend in a water pipe to hold water so sewer and/or other gases from the plumbing system do not escape into the building.

**Tread**—The horizontal part of a stair step.

**Truss**—A combination of structural members so arranged and fastened together that external loads applied at the joints will result in only direct stress in the members.

**Valley**—The depression at the meeting point of two roof slopes.

**Vault**—An arched ceiling;—a strong enclosure.

**Veneer**—An outer facing of brick, stone or other material placed on a wall for protection or decoration and not for strength.

**Venetian Window**—A window with one large fixed central pane and smaller panes at each side.

**Vent pipe**—A pipe which allows gas to escape from plumbing systems.

**Vestibule**—A small hall at the entrance of a building.

**Verge**—The edge of tiles, slates or shingles, projecting over the gable of a roof.

**Wainscoting**—The lower facing of an interior wall when different from the remainder of the wall facing.

**Water table**—A slight projection on the outside of the wall a few feet above the ground, for the purpose of shedding water.

**Weatherstripping**—Metal, wood, plastic or other material so designed that when installed at doors or windows they will retard the passage of air, water, moisture, or dust around door and window openings.

**Weep hole**—A small hole in a wall or pipe which permits water to drain off.

**Wing**—A building projection out from the main building.

**Wythe**—The partition between two chimney flues in the same stack. Also the inner and outer walls of a cavity wall.

# Appendix

## SPACE REQUIREMENTS

**Walking between two high walls (space adequate for both men and women)** — 26″

**Two people passing (figure derived; twice the space for one person to walk between two high walls)** — 52″

**Walking between high wall and 30″ high table (space adequate for both men and women)** — 26″

**Walking with elbows extended (space adequate for both men and women)** — 40″

**Reaching over obstruction, 24″ deep and 36″ high** — 78″ · 66″ · 24″ · 24″

**Reaching over obstruction, 12″ deep and 36″ high (women only)** — 74″ · 12″

**Maximum reach to back of shelf 12″ deep (women only)** — 62″ · 12″

**Using a base cabinet** — 36″

**Using a front-opening dishwasher requires 4 inches more space than using other appliances in a kitchen** — 20″ · 42″

**Using a cleaning closet** — 42″ · 48″

*Courtesy University of Illinois*

# SPACE REQUIREMENTS

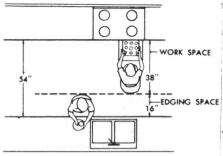

Minimum space (allowing for edging) for two people working at cabinets and appliances opposite each other (except a front-opening dishwasher)

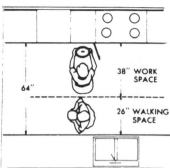

Liberal space (allowing for walking) for two people working at cabinets and appliances opposite each other (except a front-opening dishwasher)

Reaching, maximum height

Kneeling on one knee (woman only)

Using a conventional range

Using a wall oven

Man bending at a right angle

Using a refrigerator

*Courtesy University of Illinois*

## SPACE REQUIREMENTS

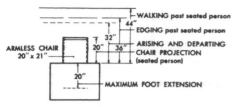

WALKING past seated person
EDGING past seated person
ARISING AND DEPARTING
CHAIR PROJECTION (seated person)
MAXIMUM FOOT EXTENSION
ARMLESS CHAIR 20" x 21"
44"
32"
20"
36"
20"

Armless chair in place at table

Rising from table, armless chair (armchair 2" more)

32"

Foot extension, knees crossed, not at table

30"

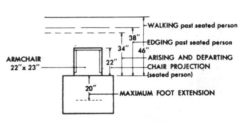

WALKING past seated person
EDGING past seated person
ARISING AND DEPARTING
CHAIR PROJECTION (seated person)
MAXIMUM FOOT EXTENSION
ARMCHAIR 22" x 23"
38"
34"
46"
22"
20"

Armchair in place at table

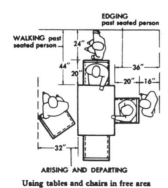

EDGING past seated person
WALKING past seated person
24"
44"
20"
36"
20"  16"
32"
ARISING AND DEPARTING

Using tables and chairs in free area

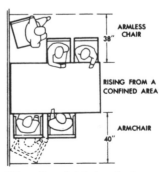

ARMLESS CHAIR
38"
RISING FROM A CONFINED AREA
ARMCHAIR
40"

Using tables and chairs in confined area

*Courtesy University of Illinois*

353

# SPACE REQUIREMENTS

**Walking past seated person**

24″

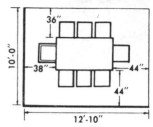

Dining areas for eight persons with free-standing table 72″ x 40″, one armchair, and seven armless chairs (calculated on basis of edging space on sides where there is not serving space, so that everyone can leave his place without disturbing others)

36″
10′-0″
38″
44″
44″
12′-10″

Serving space on one side and one end

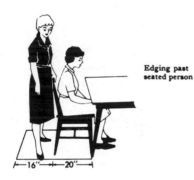

**Edging past seated person**

16″  20″

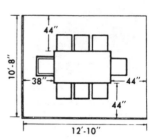

44″
10′-8″
38″
44″
44″
12′-10″

Serving space on two sides and one end

**Arising from a card table**

30″

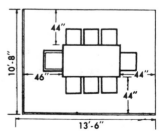

44″
10′-8″
46″
44″
44″
13′-6″

Serving space all around table

*Courtesy University of Illinois*

354

# SPACE REQUIREMENTS

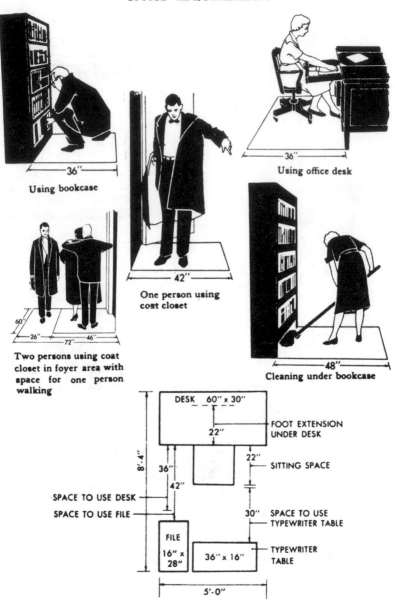

Using bookcase

One person using
coat closet

Two persons using coat
closet in foyer area with
space for one person
walking

Using office desk

Cleaning under bookcase

DESK   60" x 30"

FOOT EXTENSION
UNDER DESK

22"

SITTING SPACE

SPACE TO USE DESK

SPACE TO USE FILE

36"

42"

22"

8'-4"

30"   SPACE TO USE
TYPEWRITER TABLE

FILE
16" x
28"

36" x 16"

TYPEWRITER
TABLE

5'-0"

Parallel arrangement of office equipment

# SPACE REQUIREMENTS

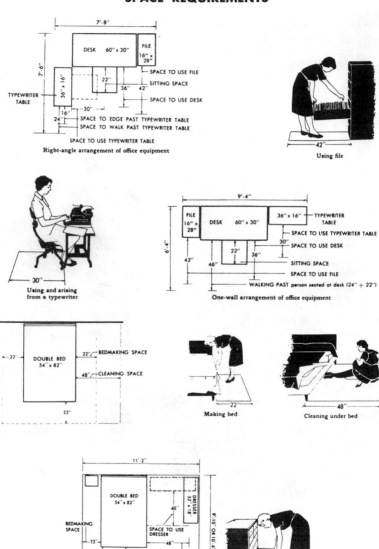

Right-angle arrangement of office equipment

Using file

Using and arising from a typewriter

One-wall arrangement of office equipment

Making bed

Cleaning under bed

Using dresser

# SPACE REQUIREMENTS

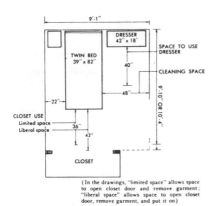

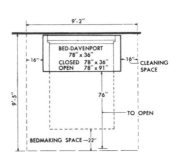

(In the drawings, "limited space" allows space
to open closet door and remove garment;
"liberal space" allows space to open closet
door, remove garment, and put it on)

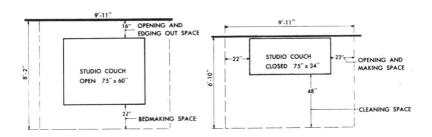

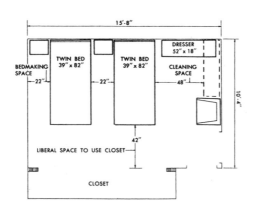

# SPACE REQUIREMENTS

Cleaning ends of bed-davenport

Making studio couch

**Opening and edging-out space
(type tested needed to be
moved out from wall to be
opened; some do not)**

Opening or making bed-davenport

Cleaning under bed-davenport
or studio couch

*Courtesy University of Illinois*

## CONCRETE REQUIREMENTS
### Footings

| | Material | | |
| SIZE | Cubic Feet Concrete Per Linear Foot | Cubic Feet Concrete Per 100 Lin. Feet | Cubic Yards Concrete Per 100 Lin. Feet |
| --- | --- | --- | --- |
| 6 x 12 | 0.50 | 50.00 | 1.9 |
| 8 x 12 | 0.67 | 66.67 | 2.5 |
| 8 x 16 | 0.89 | 88.89 | 3.3 |
| 8 x 18 | 1.00 | 100.00 | 3.7 |
| 10 x 12 | 0.83 | 83.33 | 3.1 |
| 10 x 16 | 1.11 | 111.11 | 4.1 |
| 10 x 18 | 1.25 | 125.00 | 4.6 |
| 12 x 12 | 1.00 | 100.00 | 3.7 |
| 12 x 16 | 1.33 | 133.33 | 4.9 |
| 12 x 20 | 1.67 | 166.67 | 6.1 |
| 12 x 24 | 2.00 | 200.00 | 7.4 |

### Walls

| | Material Per 100 Square Feet Wall | |
| Wall Thickness | Cubic Feet Required | Cubic Yards Required |
| --- | --- | --- |
| 4" | 33.3 | 1.24 |
| 6" | 50.0 | 1.85 |
| 8" | 66.7 | 2.47 |
| 10" | 83.3 | 3.09 |
| 12" | 100.0 | 3.70 |

### Slabs

| | Material | |
| | Per Square Foot | Square Feet from One Cubic Yard |
| Thickness | Cubic Feet of Concrete | |
| --- | --- | --- |
| 2" | 0.167 | 162 |
| 3" | 0.25 | 108 |
| 4" | 0.333 | 81 |
| 5" | 0.417 | 65 |
| 6" | 0.50 | 54 |

*Courtesy Georgia-Pacific*

# PLYWOOD INFORMATION
## Plywood Thicknesses, Spans, and Nailing Recommendations
### (Plywood Continuous Over 2 or More Spans; Grain of Face Plys Across Supports)
## Fir Plywood Floor Construction

| Application | Recommended Thickness | Maximum Spac. of Supports (C. to C.) | Nail Size and Type | Nail Spacing | |
|---|---|---|---|---|---|
| | | | | Panel Edges | Intermediate |
| Subflooring | 1/2" (a) | 16" (b) | 6d Common (c) | 6" | 10" |
| | 5/8" (a) | 20" | 8d Common (c) | 6" | 10" |
| | 3/4" (a) | 24" | 8d Common (c) | 6" | 10" |
| | 2.4.1 | 48" | 8d Ring Shank (c) | 6" | 6" |
| Underlayment | 3/8" (d) | | 6d Ring Shank or Cement Coated 8d Flathead | 6" | 8" Ea. Way |
| | 5/8" | | | | |

(a) Provide blocking at panel edges for carpet, tile, linoleum or other non-structural flooring. No blocking required for 25/32" strip flooring.
(b) If strip flooring is perpendicular to supports 1/2" can be used on 24" span.
(c) If resilient flooring is to be applied without underlayment, set nails 1/16".
(d) FHA accepts 1/4" plywood.

If supports are not well seasoned, use ring-shank nails.

## Fir Plywood Roof Sheathing

| Recommended Thickness | Max. Spacing of Supports, (C. to C.) | | | Nail Size and Type | Nail Spacing | |
|---|---|---|---|---|---|---|
| | 20 PSF | 30 PSF | 40 PSF | | Panel Edge | Intermediate |
| 5/16" | 20" (b) | 20" | 20" | 6d Common | 6" | 12" |
| 3/8" | 24" (b) | 24" | 24" | 6d Common | 6" | 12" |
| 1/2" (a) | 32" (b) | 32" | 30" | 6d Common | 6" | 12" |
| 5/8" (a) | 42" (b) | 42" | 39" | 8d Common | 6" | 12" |
| 3/4" (a) | 48" (b) | 47" | 42" | 8d Common | 6" | 12" |

(a) Provide blocking or other means of suitable edge support when span exceeds 28" for 1/2"; 32" for 5/8" and 36" for 3/4".
(b) For special case of two span continuous beams, plywood spans can be increased 6-1/2% except as roof indicated by (b) in chart.

# LUMBER INFORMATION
## Partition Stud Requirements

**Number of Feet of Lumber Required Per Sq. Ft. of Wood Stud
Partition Using 2 x 4 Studs.
Studs spaced 16" on centers, with single top and bottom plates.**

| Length Partition in Feet | No. Studs Required | Ceiling Height in Feet | | | |
|---|---|---|---|---|---|
| | | 8'-0" | 9'-0" | 10'-0" | 12'-0" |
| 2 | 3 | 1.25 | 1.167 | 1.13 | 1.13 |
| 3 | 3 | 0.833 | .812 | .80 | .80 |
| 4 | 4 | 0.833 | .812 | .80 | .80 |
| 5 | 5 | 0.833 | .812 | .80 | .80 |
| 6 | 6 | 0.833 | .812 | .80 | .80 |
| 7 | 6 | 0.833 | .75 | .75 | .80 |
| 8 | 7 | 0.75 | .75 | .75 | .70 |
| 9 | 8 | 0.75 | .75 | .75 | .70 |
| 10 | 9 | 0.75 | .75 | .75 | .70 |
| 11 | 9 | 0.75 | .70 | .70 | .67 |
| 12 | 10 | 0.75 | .70 | .70 | .67 |
| 13 | 11 | 0.75 | .70 | .70 | .67 |
| 14 | 12 | 0.75 | .70 | .70 | .67 |
| 15 | 12 | 0.70 | .70 | .70 | .67 |
| 16 | 13 | 0.70 | .70 | .70 | .67 |
| 17 | 14 | 0.70 | .70 | .70 | .67 |
| 18 | 15 | 0.70 | .70 | .67 | .67 |
| 19 | 15 | 0.70 | .70 | .67 | .67 |
| 20 | 16 | 0.70 | .70 | .67 | .67 |
| For dbl. plate, add per sq. ft. | | 0.13 | .11 | .10 | .083 |

For 2" x 8" studs, double above quantities.
For 2" x 6" studs, increase above quantities 50%.

Example: Find the number of feet of lumber, b.m. required for a stud partition 18'-0" long and 9'-0" high. This partition would contain 18 x 9 = 162 sq. ft. The table gives 0.70 ft. of lumber, b.m. per sq. ft. of partition. Multiply 162 by 0.70 equals 113.4 ft. b.m.                    *Courtesy Georgia-Pacific*

## Rafters

| | Board Feet Required 100 Square Feet Surface Area | | | Nails |
|---|---|---|---|---|
| | 12" O.C. | 16" O.C. | 24" O.C. | Per 1000 Board Feet |
| 2 x 4 | 89 | 71 | 53 | 17 |
| 2 x 6 | 129 | 102 | 75 | 12 |
| 2 x 8 | 171 | 134 | 112 | 9 |
| 2 x 10 | 212 | 197 | 121 | 7 |
| 2 x 12 | 252 | 197 | 143 | 6 |

NOTE: Includes common, hip and valley rafters, ridge boards, and collar beams.

# LUMBER INFORMATION
## Exterior Wall Studs

**(Studs Including Corner Bracing)**

| Size of Studs | Spacing on Centers | Bd. Ft. per Sq. Ft. of Area | Lbs. Nails per 1000 Bd. Ft. |
|---|---|---|---|
| 2 x 3 | 12"<br>16"<br>20"<br>24" | .83<br>.78<br>.74<br>.71 | 30 |
| 2 x 4 | 12"<br>16"<br>20"<br>24" | 1.09<br>1.05<br>.98<br>.94 | 22 |
| 2 x 6 | 12"<br>16"<br>20"<br>24" | 1.66<br>1.51<br>1.44<br>1.38 | 15 |

## Sheathing and Subflooring

**(Horizontal Application)**

| Type | Size | Bd. Ft. per Sq. Ft. of Area | Lbs. Nails Per 1000 Board feet Spacing of Framing Members | | | |
|---|---|---|---|---|---|---|
| | | | 12" | 16" | 20" | 24" |
| T & G | 1 x 4<br>1 x 6<br>1 x 8<br>1 x 10 | 1.32<br>1.23<br>1.19<br>1.17 | 66<br>43<br>32<br>37 | 52<br>33<br>24<br>29 | 44<br>28<br>21<br>24 | 36<br>23<br>17<br>20 |
| Shiplap | 1 x 4<br>1 x 6<br>1 x 8<br>1 x 10 | 1.38<br>1.26<br>1.21<br>1.18 | 69<br>44<br>32<br>37 | 55<br>34<br>25<br>29 | 46<br>29<br>21<br>25 | 38<br>24<br>17<br>20 |
| S4S | 1 x 4<br>1 x 6<br>1 x 8<br>1 x 10 | 1.19<br>1.15<br>1.15<br>1.14 | 60<br>40<br>30<br>36 | 47<br>31<br>23<br>28 | 40<br>26<br>20<br>24 | 33<br>22<br>17<br>19 |

*Courtesy Georgia-Pacific*

362

# LUMBER INFORMATION
## Wood Floor Joist Requirements
### for any Floor and Spacing

| Length of Span | Spacing of Joists | | | | | | | | | |
|---|---|---|---|---|---|---|---|---|---|---|
| | 12″ | 16″ | 20″ | 24″ | 30″ | 36″ | 42″ | 48″ | 54″ | 60″ |
| 6 | 7 | 6 | 5 | 4 | 3 | 3 | 3 | 3 | 2 | 2 |
| 7 | 8 | 6 | 5 | 5 | 4 | 4 | 3 | 3 | 3 | 2 |
| 8 | 9 | 7 | 6 | 5 | 4 | 4 | 3 | 3 | 3 | 3 |
| 9 | 10 | 8 | 6 | 6 | 5 | 4 | 4 | 3 | 3 | 3 |
| 10 | 11 | 9 | 7 | 6 | 5 | 4 | 4 | 4 | 3 | 3 |
| 11 | 12 | 9 | 8 | 7 | 5 | 5 | 4 | 4 | 3 | 3 |
| 12 | 13 | 10 | 8 | 7 | 6 | 5 | 4 | 4 | 4 | 3 |
| 13 | 14 | 11 | 9 | 8 | 6 | 5 | 5 | 4 | 4 | 4 |
| 14 | 15 | 12 | 9 | 8 | 7 | 6 | 5 | 5 | 4 | 4 |
| 15 | 16 | 12 | 10 | 9 | 7 | 6 | 5 | 5 | 4 | 4 |
| 16 | 17 | 13 | 11 | 9 | 7 | 6 | 6 | 5 | 5 | 4 |
| 17 | 18 | 14 | 11 | 10 | 8 | 7 | 6 | 5 | 5 | 4 |
| 18 | 19 | 15 | 12 | 10 | 8 | 7 | 6 | 6 | 5 | 4 |
| 19 | 20 | 15 | 12 | 11 | 9 | 7 | 6 | 6 | 5 | 5 |
| 20 | 21 | 16 | 13 | 11 | 9 | 8 | 7 | 6 | 5 | 5 |
| 21 | 22 | 17 | 14 | 12 | 9 | 8 | 7 | 6 | 6 | 5 |
| 22 | 23 | 18 | 14 | 12 | 10 | 8 | 7 | 7 | 6 | 5 |
| 23 | 24 | 18 | 15 | 13 | 10 | 9 | 8 | 7 | 6 | 6 |
| 24 | 25 | 19 | 15 | 13 | 11 | 9 | 8 | 7 | 6 | 6 |
| 25 | 26 | 20 | 16 | 14 | 11 | 9 | 8 | 7 | 7 | 6 |
| 26 | 27 | 21 | 17 | 14 | 11 | 10 | 8 | 8 | 7 | 6 |
| 27 | 28 | 21 | 17 | 15 | 12 | 10 | 9 | 8 | 7 | 6 |
| 28 | 29 | 22 | 18 | 15 | 12 | 10 | 9 | 8 | 7 | 7 |
| 29 | 30 | 23 | 18 | 16 | 13 | 11 | 9 | 8 | 7 | 7 |
| 30 | 31 | 24 | 19 | 16 | 13 | 11 | 10 | 9 | 8 | 7 |
| 31 | 32 | 24 | 20 | 17 | 13 | 11 | 10 | 9 | 8 | 7 |
| 32 | 33 | 25 | 20 | 17 | 14 | 12 | 10 | 9 | 8 | 7 |
| 33 | 34 | 26 | 21 | 18 | 14 | 12 | 10 | 9 | 8 | 8 |
| 34 | 35 | 27 | 21 | 18 | 15 | 12 | 11 | 10 | 9 | 8 |
| 35 | 36 | 27 | 22 | 19 | 15 | 13 | 11 | 10 | 9 | 8 |
| 36 | 37 | 28 | 23 | 19 | 15 | 13 | 11 | 10 | 9 | 8 |
| 37 | 38 | 29 | 23 | 20 | 16 | 13 | 12 | 10 | 9 | 8 |
| 38 | 39 | 30 | 24 | 20 | 16 | 14 | 12 | 11 | 9 | 9 |
| 39 | 40 | 30 | 24 | 21 | 17 | 14 | 12 | 11 | 10 | 9 |
| 40 | 41 | 31 | 25 | 21 | 17 | 14 | 12 | 11 | 10 | 9 |

One joist has been added to each of the above quantities to take care of extra joist required at end of span.

Add for doubling joists under all partitions.

*Courtesy Georgia-Pacific*

# LUMBER INFORMATION
## Number of Feet of Lumber B.M. Required per 100 Sq. Ft. of Surface When Used for Studs, Joists, Rafters, Wall and Floor Furring Strips, Etc.

The following table does not include any allowance for waste in cutting, doubling joists under partitions or around stair wells, extra joists at end of each span, top or bottom plates, etc. These items vary with each job. Add as required.

| Size of Lumber | 12-Inch Centers | 16-Inch Centers | 20-Inch Centers | 24-Inch Centers |
|---|---|---|---|---|
| 1 x 2 | 16-2/3 | 12-1/2 | 10 | 8-1/3 |
| 2 x 2 | 33-1/3 | 25 | 20 | 16-2/3 |
| 2 x 4 | 66-2/3 | 50 | 40 | 33-1/3 |
| 2 x 5 | 83-1/3 | 62-1/2 | 50 | 41-2/3 |
| 2 x 6 | 100 | 75 | 60 | 50 |
| 2 x 8 | 133-1/3 | 100 | 80 | 66-2/3 |
| 2 x 10 | 166-2/3 | 125 | 100 | 83-1/3 |
| 2 x 12 | 200 | 150 | 120 | 100 |
| 2 x 14 | 233-1/3 | 175 | 140 | 116-2/3 |
| 3 x 6 | 150 | 112-1/2 | 90 | 75 |
| 3 x 8 | 200 | 133-1/3 | 120 | 100 |
| 3 x 10 | 250 | 187-1/2 | 150 | 125 |
| 3 x 12 | 300 | 225 | 180 | 150 |
| 3 x 14 | 350 | 262-1/2 | 210 | 175 |

*Courtesy Georgia-Pacific*

## Furring Requirements

| Size of Strips | Spacing on Centers | Board Feet per Square Feet of Area | Lbs. Nails per 1000 Board Feet |
|---|---|---|---|
| 1 x2 | 12"<br>16"<br>20"<br>24" | .18<br>.14<br>.11<br>.10 | 55 |
| 1 x 3 | 12"<br>16"<br>20"<br>24" | .28<br>.21<br>.17<br>.14 | 37 |
| 1 x 4 | 12"<br>16"<br>20"<br>24" | .36<br>.28<br>.22<br>.20 | 30 |

*Courtesy Georgia-Pacific*

# LUMBER INFORMATION
## Square-Edged Boards

Following data is based on lumber surfaced 1 or 2 sides and 1 edge. The waste allowance shown includes width lost in dressing plus 5 percent waste in end-cutting. If laid diagonally, add 5 percent additional waste.

| Measured Size, Inches | Finished Width, Inches | Add for Waste % | Quantity Lbr. Required, Multiply Area by | Feet of Lumber Required, 100 Sq. Ft. Surface |
|---|---|---|---|---|
| 1 x 3 | 2-1/2 | 25 | 1.25 | 125 |
| 1 x 4 | 3-1/2 | 20 | 1.20 | 120 |
| 1 x 6 | 5-1/2 | 14 | 1.14 | 114 |
| 1 x 8 | 7-1/2 | 12 | 1.12 | 112 |
| 1 x 10 | 9-1/2 | 10 | 1.10 | 110 |
| 1 x 12 | 11-1/2 | 9-1/2 | 1.095 | 109-1/2 |
| 2 x 4 | 3-1/2 | 20 | 2.40 | 240 |
| 2 x 6 | 5-1/2 | 14 | 2.28 | 228 |
| 2 x 8 | 7-1/2 | 12 | 2.25 | 225 |
| 2 x 10 | 9-1/2 | 10 | 2.20 | 220 |
| 2 x 12 | 11-1/2 | 9-1/2 | 2.19 | 219 |
| 3 x 6 | 5-1/2 | 14 | 3.43 | 343 |
| 3 x 8 | 7-1/2 | 12 | 3.375 | 337-1/2 |
| 3 x 10 | 9-1/2 | 10 | 3.30 | 330 |
| 3 x 12 | 11-1/2 | 9-1/2 | 3.29 | 329 |

## Drop Siding

| | Material | | | Nails |
|---|---|---|---|---|
| | Siding for 100 Square Foot Wall | | | |
| Size | Exposed to Weather | Add for Lap | BM per 100 Sq. Ft. | Per 100 Square Feet |
| 1 x 6 | 5-1/4 | 14% | 119 | 2-1/2 Pounds |
| 1 x 8 | 7-1/4 | 10% | 115 | 2 Pounds |

NOTE: Quantities include 5% for endcutting and waste. Deduct for all openings over ten square feet.

*Courtesy Georgia-Pacific*

# LUMBER INFORMATION
## T & G, Shiplap Boards

Following data applicable on most D&M lumber. Waste allowance shown includes width loss in dressing and lapping plus 5 percent waste in end-cutting and matching.

| Measured Size, Inches | Finished Width, Inches | Add for Waste % | Quantity Lbr. Required, Multiply Area by | Feet of Lumber Required, 100 Sq. Ft. Surface |
|---|---|---|---|---|
| 1    x  2 | 1-3/8 | 50 | 1.50 | 150 |
| 1    x  2-3/4 | 2 | 42-1/2 | 1.425 | 142-1/2 |
| 1    x  3 | 2-1/4 | 38-1/3 | 1.383 | 138 |
| 1    x  4 | 3-1/4 | 28 | 1.28 | 128 |
| 1    x  6 | 5-1/4 | 20 | 1.20 | 120 |
| 1    x  8 | 7-1/4 | 16 | 1.15 | 115 |
| 1-1/4 x  3 | 2-1/4 | 38-1/3 | 1.73 | 173 |
| 1-1/4 x  4 | 3-1/4 | 28 | 1.60 | 160 |
| 1-1/4 x  6 | 5-1/4 | 20 | 1.50 | 150 |
| 1-1/2 x  3 | 2-1/4 | 38-1/3 | 2.08 | 208 |
| 1-1/2 x  4 | 3-1/4 | 28 | 1.92 | 192 |
| 1-1/2 x  6 | 5-1/4 | 20 | 1.80 | 180 |
| 2    x  4 | 3-1/4 | 28 | 2.60 | 260 |
| 2    x  6 | 5-1/4 | 20 | 2.40 | 240 |
| 2    x  8 | 7-1/4 | 16 | 2.32 | 232 |
| 2    x 10 | 9-1/4 | 13 | 2.25 | 225 |
| 2    x 12 | 11-1/4 | 12 | 2.24 | 224 |
| 3    x  6 | 5-1/4 | 20 | 3.60 | 360 |
| 3    x  8 | 7-1/4 | 16 | 3.48 | 348 |
| 3    x 10 | 9-1/4 | 13 | 3.39 | 339 |
| 3    x 12 | 11-1/4 | 12 | 3.36 | 336 |

## Built-Up Girders

| Size of Girder | Bd. Ft. per Lin. Ft. | Nails per 1000 Bd. Ft. |
|---|---|---|
| 4 x 6 | 2.15 | 53 |
| 4 x 8 | 2.85 | 40 |
| 4 x 10 | 3.58 | 32 |
| 4 x 12 | 4.28 | 26 |
| 6 x 6 | 3.21 | 43 |
| 6 x 8 | 4.28 | 32 |
| 6 x 10 | 5.35 | 26 |
| 6 x 12 | 6.42 | 22 |
| 8 x 8 | 5.71 | 30 |
| 8 x 10 | 7.13 | 24 |
| 8 x 12 | 8.56 | 20 |

## Partition Studs

(Studs including top and bottom plates)

| Size of Studs | Spacing on Centers | Bd. Ft. per Sq. Ft. of Area | Lbs. Nails per 1000 Bd. Ft. |
|---|---|---|---|
| 2 x 3 | 12" | .91 | |
|  | 16" | .83 | 25 |
|  | 24" | .76 | |
| 2 x 4 | 12" | 1.22 | |
|  | 16" | 1.12 | 19 |
|  | 24" | 1.02 | |
| 2 x 6 | 16" | 1.48 | |
|  | 24" | 1.22 | 16 |

# NAIL INFORMATION
## Standard Nail Requirements

| Description of Material | Unit of Measure | Size and Kind of Nail | Number of Nails Required | Pounds of Nails Required |
|---|---|---|---|---|
| Wood Shingles | 1,000 | 3d Common | 2,560 | 4 lbs. |
| Individual Asphalt Shingles | 100 sq.ft. | 7/8" Roofing | 848 | 4 lbs. |
| Three in One Asphalt Shingles | 100 sq. ft. | 7/8" Roofing | 320 | 1 lb. |
| Wood Lath | 1,000 | 3d Fine | 4,000 | 6 lbs. |
| Wood Lath | 1,000 | 2d Fine | 4,000 | 4 lbs. |
| Bevel or Lap Siding, 1/2" x 4" | 1,000' | 6d Coated | 2,250 | *15 lbs. |
| Bevel or Lap Siding, 1/2" x 6" | 1,000' | 6d Coated | 1,500 | *10 lbs. |
| Byrkit Lath, 1" x 6" | 1,000' | 6d Common | 2,400 | 15 lbs. |
| Drop Siding, 1" x 6" | 1,000' | 8d Common | 3,000 | 25 lbs. |
| 3/8" Hardwood Flooring | 1,000' | 4d Finish | 9,300 | 16 lbs. |
| 13/16" Hardwood Flooring | 1,000' | 8d Casing | 9,300 | 64 lbs. |
| Softwood Flooring, 1" x 3" | 1,000' | 8d Casing | 3,350 | 23 lbs. |
| Softwood Flooring, 1" x 4" | 1,000' | 8d Casing | 2,500 | 17 lbs. |
| Softwood Flooring, 1" x 6" | 1,000' | 8d Casing | 2,600 | 18 lbs. |
| Ceiling, 5/8" x 4" | 1,000' | 6d Casing | 2,250 | 10 lbs. |
| Sheathing Boards, 1" x 4" | 1,000' | 8d Common | 4,500 | 40 lbs. |
| Sheathing Boards, 1" x 6" | 1,000' | 8d Common | 3,000 | 25 lbs. |
| Sheathing Boards, 1" x 8" | 1,000' | 8d Common | 2,250 | 20 lbs. |
| Sheathing Boards, 1" x 10" | 1,000' | 8d Common | 1,800 | 15 lbs. |
| Sheathing Boards, 1" x 12" | 1,000' | 8d Common | 1,500 | 12-1/2 lbs. |
| Studding, 2" x 4" | 1,000' | 16d Common | 500 | 10 lbs. |
| Joist, 2" x 6" | 1,000' | 16d Common | 332 | 7 lbs. |
| Joist, 2" x 8" | 1,000' | 16d Common | 252 | 5 lbs. |
| Joist, 2" x 10" | 1,000' | 16d Common | 200 | 4 lbs. |
| Joist, 2" x 12" | 1,000' | 16d Common | 168 | 3-1/2 lbs. |
| Interior Trim, 5/8" thick | 1,000' | 6d Finish | 2,250 | 7 lbs. |
| Interior Trim, 3/4" Thick | 1,000' | 8d Finish | 3,000 | 14 lbs. |
| 5/8" Trim where nailed to jamb | 1,000' | 4d Finish | 2,250 | 3 lbs. |
| 1" x 2" Furring or Bridging | 1,000' | 6d Common | 2,400 | 15 lbs. |
| 1" x 1" Grounds | 1,000' | 6d Common | 4,800 | 30 lbs. |

*NOTE: Cement Coated Nails sold as two-thirds of Pound equals 1 Pound of Common Nails.

NOTE: Quantities determined for 1000 board feet of material or 100 sq. ft of Surface.

*Courtesy Georgia-Pacific*

# NAIL INFORMATION
## Aluminum Nails

| No. Nails per Lb. | Size and Type of Nail | No. Nails per Box | Coverage |
|---|---|---|---|
| 604 | 6d Wood Siding—Sinker Hd. | 575 | 500 bd. ft. 1/2"x8" Bevel Sdg. |
| 468 | 7d Wood Siding—Sinker Hd. | 575 | 500 bd. ft. 3/4"x8" Bevel Sdg. |
| 319 | 8d Wood Siding—Sinker Hd. | 575 | 500 bd. ft. 3/4"x8" Bevel Sdg. |
| 185 | 10d Wood Siding—Sinker Hd. | 290 | 250 bd. ft. 3/4"x8" Bevel Sdg. |
| 604 | 6d Wood Siding—Casing Hd. | 575 | 500 bd. ft. 1/2"x8" Bevel Sdg. |
| 468 | 7d Wood Siding—Casing Hd. | 575 | 500 bd. ft. 3/4"x8" Bevel Sdg. |
| 319 | 8d Wood Siding—Casing Hd. | 575 | 500 bd. ft. 3/4"x8" Bevel Sdg. |
| 185 | 10d Wood Siding—Casing Hd. | | |
| 1230 | 1-1/4" Asbestos Siding | 885 | 5 sqs. Asb. Sdg. Face Nailing |
| 720 | 1-3/4" Asbestos Siding | 885 | 5 sqs. Asb. Sdg. Face Nailing |
| 785 | 1-1/4" Asbestos Shingle | 885 | 5 sqs. Asb. Sdg. Conc. Nailing |
| 659 | 1-1/2" Asbestos Shingle | 885 | 5 sqs. Asb. Sdg. Conc. Nailing |
| 544 | 1-3/4" Asbestos Shingle | 885 | 5 sqs. Asb. Sdg. Conc. Nailing |
| 1300 | 1-1/4" Cedar Shake | 1680 | 3 sqs. Single Course |
| 724 | 1-3/4" Cedar Shake | 1680 | 3 sqs. Double Course |
| 1480 | 3d Cedar Shingle | 3150 | 3 sqs. with 5" Exposure |
| 1313 | 7/8" Standard Shingle | 2600 | General Purpose |
| 1009 | 3rd Standard Shingle | 2000 | Barn Battens, Joist Lining, etc. |
| 1048 | 1-3/8" Dri-Wall | 1530 | 1000 sq. ft. 3/8" Sheet Rock |
| 900 | 1-5/8" Dri-Wall | 1530 | 1000 sq. ft. 1/2" Sheet Rock |
| 988 | 1-1/8" Rock Lath | 2666 | 35 square yards |
| 939 | 1-1/4" Rock Lath | 2666 | 35 square yards |
| 725 | 1-1/2" Rock Lath | 1900 | 25 square yards |
| 495 | 2" Insulated Siding | 1680 | 5 squares |
| 295 | 2-1/2" Insulated Siding | 600 | 60 Buttress Corners |
| 663 | 7/8" Roofing | 860 | 500 sq. ft. Roll Roofing |
| 605 | 1" Roofing | 980 | 3 squares Sq. Tab. Shingles |
| 491 | 1-1/4" Roofing | 980 | 3 squares Sq. Tab. Shingles |
| 017 | 1-1/2" Roofing | 980 | 3 squares Sq. Tab. Shingles |
| 368 | 1-3/4" Roofing | 980 | 3 squares Sq. Tab. Shingles |
| 336 | 2" Roofing | 980 | 3 squares Sq. Tab. Shingles |
| 274 | 2-1/2" Roofing | 650 | 2 squares Sq. Tab. Shingles |
| 318 | 1-3/4" Roofing w/w/attached | 1050 | 10 sqs. Aluminum Roofing |
| 285 | 2" Roofing w/w/attached | 1050 | 10 sqs. Aluminum Roofing |
| 242 | 2-1/2" Roofing w/w/attached | 1050 | 10 sqs. Aluminum Roofing |

## LUMBER AND NAIL INFORMATION

### Floor Joist

| | Material | | | | Nails |
|---|---|---|---|---|---|
| | Board Feet Required for 100 Sq. Ft. of Surface Area | | | | Per 1000 Bd. Ft. |
| Size of Joist | 12" O.C. | 16" O.C. | 20" O.C. | 24" O.C. | Pounds |
| 2 x 6 | 128 | 102 | 88 | 78 | 10 |
| 2 x 8 | 171 | 136 | 117 | 103 | 8 |
| 2 x 10 | 214 | 171 | 148 | 130 | 6 |
| 2 x 12 | 256 | 205 | 177 | 156 | 5 |

### Ceiling Joist

| | Material | | | | Nails |
|---|---|---|---|---|---|
| | Board Feet Required for 100 Sq. Ft. of Surface Area | | | | Per 1000 Bd. Ft. |
| Size of Joist | 12" O.C. | 16" O.C. | 20" O.C. | 24" O.C. | Pounds |
| 2 x 4 | 78 | 59 | 48 | 42 | 19 |
| 2 x 6 | 115 | 88 | 72 | 63 | 13 |
| 2 x 8 | 153 | 117 | 96 | 84 | 9 |
| 2 x 10 | 194 | 147 | 121 | 104 | 7 |
| 2 x 12 | 230 | 176 | 144 | 126 | 6 |

### Bevel Siding

| | Material | | | Nails |
|---|---|---|---|---|
| | Siding for 100 Square Foot Wall | | | |
| Size | Exposed to Weather | Add for Lap | BM per 100 Sq. Ft. | Per 100 Square Feet |
| 1/2 x 4 | 2-3/4 | 46% | 151 | 1-1/2 Pounds |
| 1/2 x 5 | 3-3/4 | 33% | 138 | 1-1/2 Pounds |
| 1/2 x 6 | 4-3/4 | 26% | 131 | 1 Pound |
| 1/2 x 8 | 6-3/4 | 18% | 123 | 3/4 Pound |
| 5/8 x 8 | 6-3/4 | 18% | 123 | 3/4 Pound |
| 3/4 x 8 | 6-3/4 | 18% | 123 | 3/4 Pound |
| 5/8 x 10 | 8-3/4 | 14% | 119 | 1/2 Pound |
| 3/4 x 10 | 8-3/4 | 14% | 119 | 1/2 Pound |
| 3/4 x 12 | 10-3/4 | 12% | 117 | 1/2 Pound |

NOTE: Quantities include 5% for endcutting and waste. Deduct for all openings over ten square feet.

# MEASUREMENTS

## Long Measure

| | |
|---|---|
| 12 inches | 1 foot |
| 3 feet | 1 yard |
| 5-1/2 yards | 1 rod |
| 40 rods | 1 furlong |
| 8 furlongs | 1 sta. mile |
| 3 miles | 1 league |

## Square Measure

| | |
|---|---|
| 1 sq. centimeter | 0.1550 sq. in. |
| 1 sq. decimeter | 0.1076 square feet |
| 1 sq. meter | 1.196 sq. yd. |
| 1 acre | 3.954 sq. rods |
| 1 hectare | 2.47 acres |
| 1 sq. kilometer | 0.386 sq. mi. |
| 1 sq. inch | 6.452 sq. centimeters |
| 1 sq. ft. | 9.2903 sq. decimeters |
| 1 sq. yard | 0.8361 square meter |
| 1 square rod | 0.259 acre |
| 1 acre | 0.4047 hectare |
| 1 sq. mile | 2.59 sq. kilometers |
| 144 sq. inches | 1 sq. foot |
| 9 square feet | 1 square yard |
| 30-1/4 sq. yds. | 1 square rod |
| 40 sq. rods | 1 rood |
| 4 roods | 1 acre |
| 640 acres | 1 square mile |

## Surveyor's Measure

| | |
|---|---|
| 7.92 inches | 1 link |
| 25 links | 1 rod |
| 4 rods | 1 chain |
| 10 sq. chains or 160 sq. rod | 1 acre |
| 640 acres | 1 square mile |
| 36 square mi. or 6 mi. sq. | 1 township |

## Cubic Measure

| | |
|---|---|
| 1,728 cubic inches | 1 cubic foot |
| 128 cubic feet | 1 cord wood |
| 27 cubic feet | 1 cubic yard |
| 40 cubic feet | 1 ton shpg. |
| 2,150.42 cu in. | 1 standard bushel |
| 268.8 cu. in. | 1 standard gallon dry |
| 231 cu. in. | 1 standard gallon liquid |
| 1 cubic foot | about 4/5 of a bushel |
| 1 Perch | A mass 16-1/2 ft. long, 1 ft. high and 1-1/2 ft. wide, containing 24-2/3 cu. ft. |

## Approximate Metric Equivalent

| | |
|---|---|
| 1 decimeter | 4 inches |
| 1 meter | 1.1 yards |
| 1 kilometer | 5/8 of mile |
| 1 hectare | 2-1/2 acres |
| 1 stere, or cu. meter | 1/4 of a cord |
| 1 liter | 1.06 qt. liquid or 0.9 qt. dry |
| 1 hektoliter | 2.8 bushels |
| 1 kilogram | 2.2 pounds |
| 1 metric ton | 2,200 pounds |

## Metric Equivalents — Linear Measure

| | |
|---|---|
| 1 centimeter | 0.3937 in. |
| 1 decimeter | 3.937 in. or 0.328 ft. |
| 1 meter | 39.37 in. or 1.0936 yards |
| 1 dekameter | 1.9884 rods |
| 1 kilometer | 0.62137 mile |
| 1 inch | 2.54 centimeters |
| 1 foot | 3.048 decimeters |
| 1 yard | 0.9144 meter |
| 1 rod | 0.5028 dekameter |
| 1 mile | 1.6093 kilometers |

*Courtesy Georgia-Pacific*

**NOTES**

**NOTES**

**NOTES**

# The Audel® Mail Order Bookstore

Here's an opportunity to order the valuable books you may have missed before and to build your own personal, comprehensive library of Audel books. You can choose from an extensive selection of technical guides and reference books. They will provide access to the same sources the experts use, put all the answers at your fingertips, and give you the know-how to complete even the most complicated building or repairing job, in the same professional way.

## Each volume:
- **Fully illustrated**
- **Packed with up-to-date facts and figures**
- **Completely indexed for easy reference**

## APPLIANCES

### REFRIGERATION: HOME AND COMMERCIAL
Covers the whole realm of refrigeration equipment from fractional-horsepower water coolers, through domestic refrigerators to multi-ton commercial installations. 656 pages; 5½ x 8¼; hardbound. **Cat. No. 23286 Price: $9.50.**

### AIR CONDITIONING: HOME AND COMMERCIAL
A concise collection of basic information, tables, and charts for those interested in understanding, troubleshooting, and repairing home air conditioners and commercial installations. 464 pages; 5½ x 8¼; hardbound. **Cat. No. 23288 Price: $7.50.**

### HOME APPLIANCE SERVICING, 3rd Edition
A practical book for electric & gas servicemen, mechanics & dealers. Covers the principles, servicing, and repairing of home appliances. 592 pages; 5¼ x 8¼; hardbound. **Cat. No. 23214 Price: $12.95.**

### REFRIGERATION AND AIR CONDITIONING LIBRARY—2 Vols.
**Cat. No. 23305 Price: $15.95**

### OIL BURNERS, 3rd Edition
Provides complete information on all types of oil burners and associated equipment. Discusses burners—blowers—ignition transformers—electrodes—nozzles—fuel pumps—filters—Controls. Installation and maintenance are stressed. 320 pages; 5½ x 8¼; hardbound. **Cat. No. 23277 Price: $7.50.**

*Use the order coupon on the back page of this book.*

# AUTOMOTIVE

## AUTO BODY REPAIR FOR THE DO-IT-YOURSELFER

Shows how to use touch-up paint; repair chips, scratches, and dents; remove and prevent rust; care for glass, doors, locks, lids, and vinyl tops; and clean and repair upholstery. 96 pages; 8½ x 11; softcover. **Cat. No. 23238 Price: $5.95.**

## AUTOMOBILE REPAIR GUIDE, 4th Edition

A practical reference for auto mechanics, servicemen, trainees, and owners Explains theory, construction, and servicing of modern domestic motorcars. 800 pages; 5½ x 8¼; hardbound. **Cat. No. 23291 Price: $12.95.**

---

### CAN-DO TUNE-UP™ SERIES

Each book in this series comes with an audio tape cassette. Together they provide an organized set of instructions that will show you and talk you through the maintenance and tune-up procedures designed for your particular car. All books are softcover.

---

### AMERICAN MOTORS CORPORATION CARS

(The 1964 thru 1974 cars covered include: Matador. Rambler. Gremlin. and AMC Jeep (Willys).). 112 pages; 5½ x 8½; softcover. **Cat. No. 23843 Price: $7.95.**
**Cat. No. 23851** Without Cassette **Price: $4.95**

### CHRYSLER CORPORATION CARS

(The 1964 thru 1974 cars covered include: Chrysler, Dodge, and Plymouth.) 112 pages; 5½ x 8½; softcover. **Cat. No. 23825 Price $7.95.**
**Cat. No. 23846** Without Cassette **Price: $4.95**

### FORD MOTOR COMPANY CARS

(The 1954 thru 1974 cars covered include: Ford, Lincoln, and Mercury.) 112 pages; 5½ x 8½; softcover. **Cat. No. 23827 Price: $7.95.**
**Cat. No. 23848** Without Cassette **Price: $4.95**

### GENERAL MOTORS CORPORATION CARS

(The 1964 thru 1974 cars covered include: Buick, Cadillac, Chevrolet, Oldsmobile. and Pontiac.) 112 pages; 5½ x 8½; softcover. **Cat. No. 23824 Price: $7.95.**
**Cat. No. 23845** Without Cassette **Price: $4.95**

### PINTO AND VEGA CARS,

1971 thru 1974. 112 pages· 5½ x 8½; softcover. **Cat. No. 23831 Price: $7.95.**
**Cat. No. 23849** Without Cassette **Price: $4.95**

### TOYOTA AND DATSUN CARS,

1964 thru 1974. 112 pages; 5½ x 8½; softcover. **Cat. No. 23835 Price: $7.95.**
**Cat. No. 23850** Without Cassette **Price: $4.95**

### VOLKSWAGEN CARS

(The 1964 thru 1974 cars covered include: Beetle. Super Beetle. and Karmann Ghia.) 96 pages; 5½ x 8½; softcover. **Cat. No. 23826 Price: $7.95.**
**Cat. No. 23847** Without Cassette **Price: $4.95**

### AUTOMOTIVE AIR CONDITIONING

You can easily perform most all service procedures you've been paying for in the past. This book covers the systems built by the major manufacturers, even after-market installations. Contents: introduction—refrigerant—tools—air conditioning circuit—general service procedures—electrical systems—the cooling system—system diagnosis—electrical diagnosis—troubleshooting. 232 pages; 5½ x 8½; softcover. **Cat. No. 23318 Price: $5.95.**

---

*Use the order coupon on the back page of this book.*

### DIESEL ENGINE MANUAL, 3rd Edition

A practical guide covering the theory, operation, and maintenance of modern diesel engines. Explains diesel principles—valves—timing—fuel pumps—pistons and rings—cylinders—lubrication—cooling system—fuel oil and more. 480 pages; 5½ x 8¼; hardbound. **Cat. No. 23199 Price: $8.95.**

### GAS ENGINE MANUAL, 2nd Edition

A completely practical book covering the construction, operation, and repair of all types of modern gas engines. 400 pages; 5½ x 8¼; hardbound. **Cat. No. 23245 Price: $7.95.**

### OUTBOARD MOTORS & BOATING, 3rd Edition

Provides the information you need to maintain, troubleshoot, repair, and adjust all types of outboard motors. Explains the basic principles of outboard motors and the functions of the various engine parts. 464 pages; 5½ x 8¼; softcover. **Cat. No. 23279 Price: $6.95.**

# BUILDING AND MAINTENANCE

### ANSWERS ON BLUEPRINT READING, 3rd Edition

Covers all types of blueprint reading for mechanics and builders. This book reveals the secret language of blueprints, step-by-step in easy stages. 312 pages; 5½ x 8¼; hardbound. **Cat. No. 23283 Price: $6.95.**

### BUILDING A VACATION HOME

From selecting a building site to driving in the last nail, this book explains the entire process, with fully illustrated step-by-step details. Includes a complete set of drawings for a two-story vacation and/or retirement home. Softcover. 192 pages; 8½ x 11; softcover. **Cat. No. 23222 Price: $7.95.**

### BUILDING MAINTENANCE, 2nd Edition

Covers all the practical aspects of building maintenance. Painting and decorating; plumbing and pipe fitting; carpentry; heating maintenance; custodial practices and more. (A book for building owners, managers, and maintenance personnel.) 384 pages; 5½ x 8¼; hardbound. **Cat. No. 23278 Price: $7.50.**

### COMPLETE BUILDING CONSTRUCTION

At last—a *one-volume* instruction manual to show you how to construct a frame or brick building from the footings to the ridge. Build your own garage, tool shed, other outbuilding—even your own house or place of business. Building construction tells you how to lay out the building and excavation lines on the lot; how to make concrete forms and pour the footings and foundation; how to make concrete slabs, walks, and driveways; how to lay concrete block, brick and tile; how to build your own fireplace and chimney; It's one of the newest Audel books, clearly written by experts in each field and ready to help you every step of the way. 800 pages; 5½ x 8¼; hardbound. **Cat. No. 23323 Price: $19.95.**

### GARDENING & LANDSCAPING

A comprehensive guide for homeowners and for industrial, municipal, and estate groundskeepers. Gives information on proper care of annual and perennial flowers; various house plants; greenhouse design and construction; insect and rodent controls; and more. 384 pages; 5½ x 8¼; hardbound. **Cat. No. 23229 Price: $7.95.**

### CARPENTERS & BUILDERS LIBRARY, 4th Edition (4 Vols.)

A practical, illustrated trade assistant on modern construction for carpenters, builders, and all woodworkers. Explains in practical, concise language and illustrations all the principles, advances, and shortcuts based on modern practice. How to calculate various jobs. **Cat. No. 23244 Price: $24.50**

Vol. 1—Tools, steel square, saw filing, joinery cabinets. 384 pages; 5½ x 8¼; hardbound. **Cat. No. 23240 Price: $6.50.**

Vol. 2—Mathematics, plans, specifications, estimates 304 pages; 5½ x 8¼; hardbound. **Cat. No. 23241 Price: $6.50.**

Vol. 3—House and roof framing, laying out foundations. 304 pages; 5½ x 8¼; hardbound. **Cat. No. 23242 Price: $6.50.**

Vol. 4—Doors, windows, stairs, millwork, painting. 368 pages; 5½ x 8¼; hardbound. **Cat. No. 23243 Price: $6.50.**

---

*Use the order coupon on the back page of this book.*

## CARPENTRY AND BUILDING

Answers to the problems encountered in today's building trades. The actual questions asked of an architect by carpenters and builders are answered in this book. 448 pages; 5½ x 8¼; hardbound. Cat. No. 23142 Price: $7.95.

## WOOD STOVE HANDBOOK

The wood stove handbook shows how wood burned in a modern wood stove offers an immediate, practical, low-cost method of full-time or part-time home heating. The book points out that wood is plentiful, low in cost (sometimes free), and nonpolluting, especially when burned in one of the newer and more efficient stoves. In this book, you will learn about the nature of heat and its control, what happens inside and outside a stove, how to have a safe and efficient chimney, and how to install a modern wood burning stove. You will also learn about the different types of firewood and how to get it, cut it, split it, and store it. 128 pages; 8½ x 11; softcover. Cat. No. 23319 Price: $6.95.

## HEATING, VENTILATING, AND AIR CONDITIONING LIBRARY (3 Vols.)

This three-volume set covers all types of furnaces, ductwork, air conditioners, heat pumps, radiant heaters, and water heaters, including swimming-pool heating systems. Cat. No. 23227 Price: $25.50.

### Volume 1

Partial Contents: Heating Fundamentals . . . Insulation Principles . . . Heating Fuels . . . Electric Heating System . . . Furnace Fundamentals . . . Gas-Fired Furnaces . . . Oil-Fired Furnaces . . . Coal-Fired Furnaces . . . Electric Furnaces. Cat. No. 23248 Price: $8.95

### Volume 2

Partial Contents: Oil Burners . . . Gas Burners . . . Thermostats and Humidistats . . . Gas and Oil Controls . . . Pipes, Pipe Fitting, and Piping Details . . . Valves and Valve Installations. 560 pages; 5½ x 8¼; hardbound. Cat. No. 23249 Price: $8.95.

### Volume 3

Partial Contents: Radiant Heating . . . Radiators, Convectors, and Unit Heaters . . . Stoves, Fireplaces, and Chimneys . . . Water Heaters and Other Appliances . . . Central Air Conditioning Systems . . . Humidifiers and Dehumidifiers. 544 pages; 5½ x 8¼; hardbound. Cat. No. 23250 Price: $8.95.

## HOME MAINTENANCE AND REPAIR: Walls, Ceilings, and Floors

Easy-to-follow instructions for sprucing up and repairing the walls, ceiling, and floors of your home. Covers nail pops, plaster repair, painting, paneling, ceiling and bathroom tile, and sound control. 80 pages; 8½ x 11; softcover. Cat. No. 23281 Price: $5.95.

## HOME PLUMBING HANDBOOK , 2nd Edition

A complete guide to home plumbing repair and installation. 200 pages; 8½ x 11; softcover. Cat. No. 23321 Price: $6.95.

## HOME WORKSHOP & TOOL HANDY BOOK

Tells how to set up your own home workshop (basement, garage, or spare room) and explains the various hand and power tools (when, where, and how to use them). 464 pages; 5½ x 8¼; hardbound. Cat. No. 23208 Price $6.50.

## MASONS AND BUILDERS LIBRARY—2 Vols.

A practical, illustrated trade assistant on modern construction for bricklayers, stonemasons, cement workers, plasterers, and tile setters. Explains all the principles, advances, and shortcuts based on modern practice—including how to figure and calculate various jobs. Cat. No. 23185 Price: $13.95

Vol. 1—Concrete, Block, Tile, Terrazzo. 368 pages; 5½ x 8¼; hardbound. Cat. No. 23182 Price: $7.50.

Vol. 2—Bricklaying, Plastering, Rock Masonry, Clay Tile. 384 pages; 5½ x 8¼; hardbound. Cat. No. 23183 Price: $7.50.

---

*Use the order coupon on the back page of this book.*

## PLUMBERS AND PIPE FITTERS LIBRARY—3 Vols.

A practical, illustrated trade assistant and reference for master plumbers, journeymen and apprentice pipe fitters, gas fitters and helpers, builders, contractors, and engineers. Explains in simple language, illustrations, diagrams, charts, graphs, and pictures, the principles of modern plumbing and pipe-fitting practices. **Cat. No. 23255 Price $19.95**

Vol. 1—Materials, tools, roughing-in. 320 pages; 5½ x 8¼; hardbound. **Cat. No. 23256 Price: $6.95.**

Vol. 2—Welding, heating, air-conditioning. 384 pages; 5½ x 8¼; hardbound. **Cat. No. 23257 Price: $6.95.**

Vol. 3—Water supply, drainage, calculations. 272 pages; 5½ x 8¼; hardbound. **Cat. No. 23258 Price: $6.95.**

## PLUMBERS HANDBOOK

A pocket manual providing reference material for plumbers and/or pipe fitters. General information sections contain data on cast-iron fittings, copper drainage fittings, plastic pipe, and repair of fixtures. 288 pages; 4 x 6; softcover. **Cat. No. 23339 Price: $5.95.**

## QUESTIONS AND ANSWERS FOR PLUMBERS EXAMINATIONS,

### 2nd Edition

Answers plumbers' questions about types of fixtures to use, size of pipe to install, design of systems, size and location of septic tank systems, and procedures used in installing material. 256 pages; 5½ x 8¼; softcover. **Cat. No. 23285 Price: $5.50.**

## TREE CARE MANUAL

The conscientious gardener's guide to healthy, beautiful trees. Covers planting, grafting, fertilizing, pruning, and spraying. Tells how to cope with insects, plant diseases, and environmental damage. 224 pages; 8½ x 11; softcover. **Cat. No. 23280 Price: $8.95.**

## UPHOLSTERING

Upholstering is explained for the average householder and apprentice upholsterer. From repairing and regluing of the bare frame, to the final sewing or tacking, for antiques and most modern pieces, this book covers it all. 400 pages; 5½ x 8¼; hardbound. **Cat. No. 23189 Price: $7.95.**

## WOOD FURNITURE: Finishing, Refinishing, Repairing

Presents the fundamentals of furniture repair for both veneer and solid wood. Gives complete instructions on refinishing procedures, which includes stripping the old finish, sanding, selecting the finish and using wood fillers. 352 pages; 5½ x 8¼; hardbound. **Cat. No. 23216 Price: $8.50.**

# ELECTRICITY/ELECTRONICS

## ELECTRICAL LIBRARY

If you are a student of electricity or a practicing electrician, here is a very important and helpful library you should consider owning. You can learn the basics of electricity, study electric motors and wiring diagrams, learn how to interpret the NEC, and prepare for the electrician's examination by using these books. **Cat. No. 23324 Price: $43.95.**

Electric Motors, 3rd Edition. 528 pages; 5½ x 8¼; hardbound. **Cat. No. 23264 Price: $8.95.**

Guide to the 1978 National Electrical Code. 672 pages; 5½ x 8¼; hardbound. **Cat. No. 23308 Price: $9.95.**

House Wiring, 4th Edition. 256 pages; 5½ x 8¼; hardbound. **Cat. No. 23315 Price: $6.95.**

Practical Electricity, 3rd Edition. 496 pages; 5½ x 8¼; hardbound. **Cat. No. 23218 Price: $8.95**

Questions and Answers for Electricians Examinations, 6th Edition. 288 pages; 5½ x 8¼; hardbound. **Cat. No. 23307 Price: $6.95.**

Wiring Diagrams for Light and Power, 3rd Edition. 400 pages; 5½ x 8¼; hardbound. **Cat. No. 23232 Price: $6.95.**

## ELECTRICAL COURSE FOR APPRENTICES AND JOURNEYMEN

A study course for apprentice or journeymen electricians. Covers electrical theory and its applications. 448 pages; 5½ x 8¼; hardbound. **Cat. No. 23209 Price: $7.95**

*Use the order coupon on the back page of this book.*

### RADIOMANS GUIDE, 4th Edition

Contains the latest information on radio and electronics from the basics through transistors. 480 pages; 5½ x 8¼; hardbound. **Cat. No. 23259 Price: $7.50.**

### TELEVISION SERVICE MANUAL, 4th Edition

Provides the practical information necessary for accurate diagnosis and repair of both black-and-white and color television receivers. 512 pages; 5½ x 8¼; hardbound. **Cat. No. 23247 Price: $8.95.**

# ENGINEERS/MECHANICS/ MACHINISTS

### MACHINISTS LIBRARY, 2nd Edition

Covers modern machine-shop practice. Tells how to set up and operate lathes, screw and milling machines, shapers, drill presses, and all other machine tools. A complete reference library. **Cat. No. 23300 Price: $23.00**

Vol. 1—Basic Machine Shop. 352 pages; 5½ x 8¼; hardbound. **Cat. No. 23301 Price: $7.95.**

Vol. 2—Machine Shop. 480 pages; 5½ x 8¼; hardbound. **Cat. No. 23302 Price: $7.95.**

Vol. 3—Toolmakers Handy Book. 400 pages; 5½ x 8¼; hardbound. **Cat. No. 23303 Price: $7.95.**

### MECHANICAL TRADES POCKET MANUAL

Provides practical reference material for mechanical tradesmen. This handbook covers methods, tools, equipment, procedures, and much more. 256 pages; 4 x 6; softcover. **Cat. No. 23215 Price: $4.50.**

### MILLWRIGHTS AND MECHANICS GUIDE, 2nd Edition

Practical information on plant installation, operation, and maintenance for millwrights, mechanics, maintenance men, erectors, riggers, foremen, inspectors, and superintendents. 960 pages; 5½ x 8¼; hardbound. **Cat. No. 23201 Price: $11.95.**

### POWER PLANT ENGINEERS GUIDE, 2nd Edition

The complete steam or diesel power-plant engineer's library. 816 pages; 5½ x 8¼; hardbound. **Cat. No. 23220 Price: $12.95.**

### QUESTIONS AND ANSWERS FOR ENGINEERS AND FIREMANS EXAMINATIONS, 3RD EDITION

Presents both legitimate and "catch" questions with answers that may appear on examinations for engineers and firemans licenses for stationary, marine, and combustion engines. 496 pages; 5½ x 8¼; hardbound. **Cat. No. 23327 Price $8.95.**

### WELDERS GUIDE, 2nd Edition

This new edition is a practical and concise manual on the theory, practical operation, and maintenance of all welding machines. Fully covers both electric and oxy-gas welding. 928 pages; 5½ x 8¼; hardbound. **Cat. No. 23202 Price: $11.95.**

### WELDER/FITTERS GUIDE

Provides basic training and instruction for those wishing to become welder/fitters. Step-by-step learning sequences are presented from learning about basic tools and aids used in weldment assembly, through simple work practices, to actual fabrication of weldments. 160 pages· 8½ x 11; softcover; **Cat. No. 23325 Price: $7.95.**

---

*Use the order coupon on the back page of this book.*

# FLUID POWER

### PNEUMATICS AND HYDRAULICS, 3rd Edition

Fully discusses installation, operation, and maintenance of both HYDRAULIC AND PNEUMATIC (air) devices. 496 pages; 5½ x 8¼; hardbound. **Cat. No. 23237 Price: $8.50.**

### PUMPS, 3rd Edition

A detailed book on all types of pumps from the old-fashioned kitchen variety to the most modern types. Covers construction, application, installation, and troubleshooting. 480 pages; 5½ x 8¼; hardbound. **Cat. No. 23292 Price: $8.95.**

### HYDRAULICS FOR OFF-THE-ROAD EQUIPMENT

Everything you need to know from basic hydraulics to troubleshooting hydraulic systems on off-the-road equipment. Heavy-equipment operators, farmers, fork-lift owners and operators, mechanics—all need this practical, fully illustrated manual. 272 pages; 5½ x 8¼; hardbound. **Cat. No. 23306 Price: $6.95.**

# HOBBY

### COMPLETE COURSE IN STAINED GLASS

Written by an outstanding artist in the field of stained glass, this book is dedicated to all who love the beauty of the art. Ten complete lessons describe the required materials, how to obtain them, and explicit directions for making several stained glass projects. 80 pages; 8½ x 11; softbound. **Cat. No. 23287   Price: $4.95**

## BUILD YOUR OWN AUDEL DO-IT-YOURSELF LIBRARY AT HOME!

Use the handy order coupon today to gain the valuable information you need in all the areas that once required a repairman. Save money and have fun while you learn to service your own air conditioner, automobile, and plumbing. Do your own professional carpentry, masonry, and wood furniture refinishing and repair. Build your own security systems. Find out how to repair your TV or Hi-Fi. Learn landscaping, upholstery, electronics and much, much more.

# HERE'S HOW TO ORDER

1. Enter the correct catalog number(s) of the book(s) you want in the space(s) provided.

2. Print your name, address, city, state and zip code, clearly.

3. Detach the order coupon below and mail today to:

**Theodore Audel & Company**
4300 West 62nd Street
Indianapolis, Indiana 46206
**ATTENTION: ORDER DEPT.**

All prices are subject to change without notice.

------------------------------------------------------------

# ORDER COUPON

Please rush the following book(s).

| | | | | | |
|---|---|---|---|---|---|

Write book catalog num-
bers at left.
(Numbers are listed with
titles.)

NAME _____

ADDRESS _____

CITY _____ STATE _____ ZIP _____

☐ Payment Enclosed _____
   (No Shipping and    Total
   Handling Charge)

☐ Bill Me (Shipping and Handling Charge will be added)

Add local sales tax where applicable.

Litho in U.S.A.

# HERE'S HOW TO ORDER

Select the Audel book(s) you want, fill in the order card below, detach and mail today. Send no money now. You'll have 15 days to examine the books in the comfort of your own home. If not completely satisfied, simply return your order and owe nothing.

If you decide to keep the books, we will bill you for the total amount, plus a small charge for shipping and handling.

**1.** Enter the correct catalog number(s) of the book(s) you want in the space(s) provided.

**2.** Print your name, address, city, state and zip code, clearly.

**3.** Detach the order card below and mail today. No postage is required.

*Detach postage-free order card on perforated line*

---

## FREE TRIAL ORDER CARD

☐ Please rush the following book(s) for my free trial. I understand if I'm not completely satisfied, I may return my order within 15 days and owe nothing. Otherwise, you will bill me for the total amount plus a small postage & handling charge.

Write book catalog numbers at right.

(Numbers are listed with titles)

NAME_____

ADDRESS_____

CITY_____STATE_____ZIP_____

☐ Save postage & handling costs. Full payment enclosed (Plus sales tax, if any.)

Cash must accompany orders under $5.00.
Money-Back guarantee still applies.

333

## DETACH POSTAGE-PAID REPLY CARD BELOW AND MAIL TODAY!

Just select your books, enter the code numbers on the order card, fill out your name and address, and mail. There's no need to send money.

**15-Day Free Trial On All Books . . .**